Darwinian Fairytales

DAVID STOVE

Darwinian Fairytales

SELFISH GENES, ERRORS OF HEREDITY, AND OTHER FABLES OF EVOLUTION

Introduction by
Roger Kimball

ENCOUNTER BOOKS
New York

 ENCOUNTER BOOKS

Paperback edition published in 2007 by Encounter Books, an activity of Encounter for Culture and Education, Inc., a nonprofit, tax exempt corporation.

Encounter Books website address: www.encounterbooks.com

Manufactured in the United States and printed on acid-free paper.

The paper used in this publication meets the minimum requirements of ANSI/NISO Z39.48-1992 (R 1997) (*Permanence of Paper*).

Paperback edition: ISBN 1-59403-200-9

The Library of Congress has catalogued the hardcover edition as follows:

Stove, D.C. (David Charles).
  Darwinian fairytales/ David Stove.
      p. cm.
    Includes index.
    ISBN 1-59403-140-1 (alk. paper)
      1. Social Darwinism. 2. Evolution (Biology)

R726 .S576 2005
303.4 22-dc 22

2005033994

10 9 8 7 6 5 4 3 2

Contents

Introduction
by Roger Kimball

C HARLES DARWIN is back in the news. As I write, his image
gazes out from the cover of *Newsweek*, which duly reports
on the controversy over "intelligent design," an old, religiously-
based challenge to Darwinian theory dressed up in (slightly) new
rhetorical clothing. (See Essay 10 below for some of the classic
arguments for and against "intelligent design" theory). A couple
of months ago, in its issue for August 22 & 29, 2005, *The New
Republic* devoted *its* cover story to "Unintelligent Design," a
long, blistering attack on the movement by Jerry Coyne, a
biologist at the University of Chicago.

In my view, Coyne leaves the claims of the IDers in tatters, but
that is not the end of the story. What's it all about? Are we reliv-
ing the Scopes Monkey Trial, with the forces of darkness, super-
stition, and dogmatic religion arrayed on one side against the
champions of enlightenment and truth lined up on the other?

That's what many of my friends say (yours, too, probably, if
you went to a good school).

In fact, there is a lot of fog, not to say bad faith, on both sides
of the debate—or whatever we should call the current, self-
righteous exchange of name-calling.

There is also an extraordinary amount of dogmatism—as
much, interestingly, on the side of the Darwinians as on the side

of their "fundamentalist" opponents. I remember leaving a party a few years ago with a friend, an eminent biologist, whom I knew admired some of David Stove's other writings. I was in the midst of reading *Darwinian Fairytales* for the first time. At the coat check, I retrieved the book and gave it to him to leaf through. I looked on with concern as the smile on his face faded and a frown appeared. "But this is terrible!" he said, turning the pages, "this is awful!"

Since then I have noted with amusement how sensitive to criticism the Darwinian faithful are. Any hint of a shadow of dissent and they rush for the garlic, the wooden stake, and a signed copy of *On the Origin of Species*. I think I understand the psychology of the response. They are terrified lest acknowledging the strength of this or that criticism start them down the road toward creationism and teaching the Book of Genesis in Biology 101.

Where does the truth lie? Are the main doctrines of Darwinian teaching as impregnable and well established as proponents claim? One good place to start for an answer is with this book, *Darwinian Fairytales*. It was the last work of the Australian philosopher David Stove. By the time of his death in 1994, Stove had earned a distinguished place for himself in the pantheon of intellectual demolition experts. His targets, one admirer wrote, are many: "the Enlightenment, feminism, Freud, the idea of progress, leftish views of all kinds, Marx, . . . metaphysics, modern architecture and art, philosophical idealism, [Karl] Popper, religion, semiotics, Stravinsky and Sweden. . . . Also, anything beginning with 'soc' (even Socrates got a serve or two)."

High among Stove's antipathies was irrationality in the philosophy of science, which he detected not only in overt irrationalists such as Paul "Anything Goes" Feyerabend but also in more seemingly respectable figures such as Thomas "Mr. Paradigm Shift" Kuhn. If you believe (as your teachers doubtless told you) that Kuhn or Karl Popper was a friend of honest scientific inquiry, read Stove's book *Scientific Irrationalism: Origins of a Postmodern Cult* (Transaction): I guarantee that it will change your mind.

A century ago, William James wrote a book called *Varieties of Religious Experience*. Had Stove lived long enough, he might have written something called *Varieties of Irrational Experience*. His hilarious but disturbing essay "What is Wrong with Our Thoughts?," reprinted in *The Plato Cult and Other Philosophical Follies* (Blackwell), would make a splendid introduction to such a work.

I want to pause to emphasize the hilarity. Stove was one of the greatest philosophical stylists of our, perhaps of any, time. That is a large claim, I know, but don't take my word for it: plunge on. You will see why one commentator wrote that reading David Stove is like watching Fred Astaire dance: the elegance, the seeming effortlessness, are breathtaking demonstrations of consummate artistry.

Stove's skeptical cast of mind (David Hume was his philosophical hero) naturally endeared him to the enemies of cant, the exposers of naked emperors, the puncturers of academic gas bags, poseurs, and charlatans.

How surprising, then, that David Stove should turn out to be an ardent anti-Darwinian. Wasn't Darwin on the side of all us Enlightened, no-nonsense, scientifically educated folk? If David Stove criticizes Darwin's theories, doesn't that make him an irrationalist, an ally of those school boards in Kansas that (or so we are told) want to replace science with scripture?

No, it doesn't. For one thing, Stove is not a creationist or proponent of "intelligent design"; indeed, he is careful to point out that he is "of no religion." Moreover, Stove admires Darwin greatly as a thinker, placing him at the top of his personal pantheon, along with Shakespeare, Purcell, Newton, and Hume. Stove furthermore believes that it is "overwhelmingly probable" that our species evolved from some other and that "natural selection is probably the cause which is principally responsible for the coming into existence of new species from old ones." Indeed, he believes that

the Darwinian explanation of evolution is a very good one as far as it goes, and it has turned out to go an extremely long way. Its explanatory power, even in 1859 [when *On the Origin of Species* was published], was visibly very great, but it has turned out to be far greater than anyone then could have realized. And then, in the 1930s, the Darwinian theory received further accessions of explanatory strength through its confluence or synthesis with the new knowledge of genetics. And this "new synthesis," or "neo-Darwinism," has been itself growing rapidly in explanatory power ever since.

At the same time, though, Stove maintains that "Darwinism says many things, especially about our species, which are too obviously false to be believed by any educated person; or at least by an educated person who retains any capacity at all for critical thought."

Examples? Here are a few: that "every single organic being around us may be said to be striving to the utmost to increase its numbers"; that "of the many individuals of any species which are periodically born, but a small number can survive"; that it is to a mother's "advantage" that her child should be adopted by another woman; that "no one is prepared to sacrifice his life for any single person, but . . . everyone will sacrifice it for more than two brothers, or four half-brothers, or eight first cousins"; that "any variation in the least degree injurious [to a species] would be rigidly destroyed."

All of these quotations are from Darwin or his orthodox disciples. A moment's reflection shows that none is even remotely true, at least of human beings. Take the last named: that anything in the least injurious to a species would be "rigidly destroyed" by natural selection. What about abortion, adoption, fondness for alcohol, anal intercourse, or asceticism, just to start with the "A"s? As Stove notes, "each of these characteristics [tends] to shorten our lives, or to lessen the number of children we have, or both." Are any on the way to being rigidly destroyed?

Again, if Darwin's theory of evolution were true, "there would be in every species a constant and ruthless competition to survive:

a competition in which only a few in any generation can be winners. But it is perfectly obvious that human life is not like that, however it may be with other species." Priests, hospitals, governments, old-age homes, charities, police: these are a few of the things whose existence contradicts Darwin's theory.

Some of Darwinism's defenders respond by arguing that although human life may not *now* exhibit the brutal struggle for subsistence that Darwin's theory postulates, it *once did*.

> In the olden days (this story goes), human populations always did press relentlessly on their supply of food, and thereby brought about constant competition for survival among the too-numerous competitors, and hence natural selection of those organisms which were best fitted to succeed in the struggle for life. That is, human life was exactly as Darwin's book had said that all life is. But our species (the story goes on) escaped long ago from the brutal *régime* of natural selection. We developed a thousand forms of attachment, loyalty, cooperation, and unforced subordination, every one of them quite incompatible with a constant and merciless competition to survive.

This is what Stove calls the "Cave Man" attempt to solve "Darwinism's Dilemma." (The other attempts he calls the "Hard Man" and the "Soft Man" gambits—see Essay 1 below). But the problem is that Darwin's theory is not meant to be something that was true yesterday but not today. It claims to be, as Stove puts it, "a universal generalization about all terrestrial species at any time." And this means that "if the theory says something which is not true *now* of our species (or another), then it is not true—finish." Stove writes:

> If Darwin's theory of evolution is true, no species can *ever* escape from the process of natural selection. His theory is that two universal and permanent tendencies of all species of organisms—the tendency to increase in numbers up to the limit that the food supply allows, and the tendency to vary in a heritable way—are

together sufficient to bring about in any species universal and permanent competition for survival, and therefore universal and permanent natural selection among the competitors.

But this is clearly not true of our species now. Nor, Stove points out, can it *ever* have been true of our species. "It may be possible, for all I know, that a population of pines or cod should exist with no cooperative as distinct from competitive relations among its members. But no tribe of humans could possibly exist on those terms. Such a tribe could not even raise a second generation: the helplessness of the human young is too extreme and prolonged."

Stove shows in unanswerable detail that, despite its enormous explanatory power regarding "cods, pines, flies," etc., Darwin's theory of evolution is "a ridiculous slander on human beings." He is particularly good at exposing the "amazingly arrogant habit of Darwinians" of "*blaming the fact, instead of blaming their theory*" when they encounter contrary biological facts. Doctrinaire Darwinists have an answer for *everything*, always a bad sign in science, since it means that mere facts can never prove them wrong. Does it regularly happen that increasing prosperity leads to lower birth rates? And does this directly contradict Darwinian theory? No problem, just announce that the birth rates in such cases are somehow "inverted," evidence of a "biological mistake."

Even more amusing is to watch a sociobiologist tie himself in knots trying to explain—or explain away—the phenomenon of altruism among human beings. The latest wheeze is something called "inclusive fitness" (see Essay 8 below), which attempts to deal with the "problem" of altruism by transforming it into a surreptitious form of selfishness. Darwinian theory says that *really*, deep down, altruism cannot exist: we're all engaged in a war for survival, after all. And yet, as Stove points out, our species "is sharply distinguished from all other animals by being in fact *hopelessly addicted to* altruism. It will be time to think otherwise when, and not before, adult wolves or kookaburras or rats pool their resources in order to relieve the illness, or improvidence, or ignorance, of conspecifics to whom they are unrelated" (see Essay 6 below).

Stove is also very good at exposing the mind-boggling claims of sociobiology—a.k.a. evolutionary psychology—a school of neo-Darwinism whose fundamental tenet is that an organism is epiphenomenal to its genes: that a human being, for example, is nothing more than a puppet manipulated by his genetic makeup. If this seems like an exaggeration, consider the statement by the eminent sociobiologist E. O. Wilson that "an organism is only DNA's way of making more DNA." It is worth pausing to ponder the implications of that adverb "only."

Or consider Richard Dawkins, another eminent sociobiologist and author of *The Selfish Gene*, a hugely popular book whose basic message is that "we are . . . robot-vehicles blindly programmed to preserve the selfish molecules known as genes." (Yes, he really says this.) Of course, as Stove points out, "genes can no more be selfish than they can be (say) supercilious, or stupid." The popularity of Dawkins's book lies in the powerful appeal that puppet-theories of human behavior always exercise on those who combine cynicism with credulousness; but genetic puppet theories are no more credible than those propounded by Freudians, Marxists, or astrologers.

In the end, Stove's discussion of Darwinian theory shows that, when it comes to the species *H. sapiens*, Darwinism "is a mere festering mass of errors." It can tell you "lots of truths about plants, flies, fish, etc., and interesting truths, too. . . . [But] if it is *human* life that you would most like to know about and to understand, then a good library can be begun by leaving out Darwinism, from 1859 to the present hour." It is not a pretty picture that Stove paints; but then the exhibition of gross error widely accepted is never a comely sight.

Preface

This is an anti-Darwinism book. It is written both against the Darwinism of Darwin and his nineteenth-century disciples, and against the Darwinism of such influential twentieth-century Darwinians as G. C. Williams and W. D. Hamilton and their disciples. My object is to show that Darwinism is not true: not true, at any rate, of *our* species. If it is true, or near enough true, of sponges, snakes, flies, or whatever, I do not mind that. What I do mind is, its being supposed to be true of man.

But having said that, I had better add at once that I am not a "creationist," or even a Christian. In fact I am of no religion. It seems just as obvious to me as it does to any Darwinian, that the species to which I belong is a certain species of land-mammal. And it seems just as overwhelmingly probable to me as it does to any Darwinian that our species has evolved from some other animals.

I do not even deny that natural selection is probably the cause which is principally responsible for the coming into existence of new species from old ones. I do deny that natural selection is going on *within* our species *now*, and that it ever went on in our species, at any time of which anything is known. But I say nothing at all in the book about how our species came to be the kind of thing it is, or what kind of antecedents it evolved from. Such questions strike me, in fact, as overwhelmingly uninteresting: like the questions (say) where the Toltecs came from, or the Hittites, and how they came. They came, like our species itself, from somewhere, and they came somehow. The details do not matter, except to specialists. What does matter is to see our species rightly, as it now is, and as it is known historically to have been: and in particular, not to be imposed upon by the ludicrously false portrayals which Darwinians give of the past, and even of the present, of our species.

I should also say here that I have no professional qualifications of any kind for writing about Darwinism. I am not a biologist: merely a former professional philosopher, who happens to have both 40 odd years' acquaintance with Darwinian literature and a strong distaste for ridiculous slanders on our species. These are evidently not ideal qualifications for criticizing Darwinian views of man. But on the other hand, Darwinism is not yet so arcane a branch of science that criticism of it by an outsider can be automatically assumed to be incompetent.

I have called the eleven parts which make up the book "essays" rather than "chapters," because I at first intended them to be quite independent of one another, and able to be read in any order. As things have turned out, however, the essays, though largely free-standing, are not as independent of one another as I had intended. They are probably best read in the order in which they are printed here.

Darwinian Fairytales

Essay 1
Darwinism's Dilemma

... in the state of nature ... [human] life was a continual free fight.
—T. H. Huxley, *Evolution and Ethics*

I F DARWIN'S THEORY of evolution were true, there would be in every species a constant and ruthless competition to survive: a competition in which only a few in any generation can be winners. But it is perfectly obvious that human life is not like that, however it may be with other species.

This inconsistency, between Darwin's theory and the facts of human life, is what I mean by "Darwinism's Dilemma." The inconsistency is so very obvious that no Darwinian has ever been altogether unconscious of it. There have been, accordingly, very many attempts by Darwinians to wriggle out of the dilemma. But the inconsistency is just too simple and direct to *be* wriggled out of, and all these attempts are conspicuously unsuccessful. They are not uninstructive, though, or unamusing.

The attempts to escape from Darwinism's dilemma all fall into one or other of three types. These can be usefully labelled "the Cave Man way out," "the Hard Man," and "the Soft Man." All three types are hardy perennials, and have been with us, in one version or another, ever since Darwin published *The Origin of Species* in 1859.

What I call the Cave Man way out is this: you admit that human life is not now what it would be if Darwin's theory were true, but also insist that *it used to be* like that.

In the olden days (this story goes), human populations always did press relentlessly on their supply of food, and thereby brought about constant competition for survival among the too-numerous competitors, and hence natural selection of those organisms which were best fitted to succeed in the struggle for life. That is, human life was exactly as Darwin's book had said that all life is. But our species (the story goes on) escaped long ago from the brutal *régime* of natural selection. We developed a thousand forms of attachment, loyalty, cooperation, and un-forced subordination, every one of them quite incompatible with a constant and merciless competition to survive. We have now had for a very long time, at least locally, religions, moralities, laws or customs, respect for life and property, rules of in-heritance, specialized social orders, distinctions of rank, and standing provisions for external defense, internal police, educa-tion and health. Even at our lowest ebb we still have ties of blood, and ties of marriage: two things which are quite as in-compatible with a universal competition to survive as are, for example, a medical profession, a priesthood, or a state.

This Cave Man story, however implausible, is at any rate not inconsistent with itself. But the combination of it with Darwin's theory of evolution *is* inconsistent. That theory is a universal generalization about all terrestrial species at any time. Hence, if the theory says something which is not true *now* of our species (or another), then it is not true—finish. In short, the Cave Man way out of Darwinism's dilemma is in reality no way out at all: it is self-contradictory.

If Darwin's theory of evolution is true, no species can *ever* es-cape from the process of natural selection. His theory is that two universal and permanent tendencies of all species of organisms— the tendency to increase in numbers up to the limit that the food supply allows, and the tendency to vary in a heritable way—are together sufficient to bring about in any species universal and permanent competition for survival, and therefore universal and permanent natural selection among the competitors.

So the "modern" part of this way out of Darwin's dilemma is inconsistent with Darwinism. But the Cave Man part of it is also utterly incredible in itself. It may be possible, for all I know, that a population of pines or cod should exist with no cooperative as distinct from competitive relations among its members. But no tribe of humans could possibly exist on those terms. Such a tribe could not even raise a second generation: the helplessness of the human young is too extreme and prolonged. So if you ever read a report (as one sometimes does) of the existence of an on-going tribe of just this kind, you should confidently conclude that the reporter is mistaken or lying or both.

Even if such a tribe *could* somehow continue in existence, it is extremely difficult to imagine how our species, as we now know it to be, could ever have graduated from so very hard a school. We need to remember how severe the rule of natural selection is, and what it means to say that a species is subject to it. It means, among other things, that of all the rabbits, flies, cod, pines, etc., that are born, the enormous majority *must* suffer early death; and it means no less of our species. How *could* we have escaped from this set up, supposing we once were in it? Please don't say that a god came down, and pointed out to Darwinian Cave Men a better way; or that the Cave Men themselves got together and adopted a Social Contract (with a Department of Family Planning). Either of those explanations is logically possible, of course, but they are just too improbable to be worth talking about. Yet some explanation, of the same order of improbability, seems to be required, if we once allow ourselves to believe that though we are not subject now to natural selection, we used to be.

The Cave Man way out, despite its absurdity, is easily the most popular of the three ways of trying to get out of Darwinism's dilemma. It has been progressively permeating popular thought for nearly one hundred and fifty years. By now it is enshrined in a thousand cartoons and comic-strips, and it is as immovable as Christmas. But we should not infer from this that it lacks high scientific authorities in its favor. Quite the contrary, Cave Man has been all along, and still is, the preferred way out of Darwinism's dilemma among the learned, as well as among the vulgar.

Darwinism in its early decades had an urgent need for an able and energetic PR man. Darwin himself had little talent for that kind of work, and even less taste for it. But he found in T. H. Huxley someone who had both the talent and the taste in plenty. Huxley came to be known as "Darwin's bulldog," and by thirty years of invaluable service as a defender of Darwinism against all comers, he deserved it. And he provides an unusually explicit example of a high scientific authority who takes the Cave Man way out.

Huxley knew perfectly well, of course, since he was not a madman, that human life in England in his own time did not bear any resemblance to a constant and ruthless struggle to survive. Why, life was not like *that* even among the savages of New Guinea—nay, even in Sydney—as he found when he was in these parts in the late 1840s, as a surgeon on board H.M.S. *Rattlesnake*. Did these facts make him doubt, when he became a Darwinian about ten years later, the reality of Darwin's "struggle for life," at least in the case of humans? Of course not. They only made him think that, while of course there must have *been* a stage of Darwinian competition in human history, it must also have ended long ago.

BUT IN THOSE distant times, Huxley informs us, human beings lived in "nature," or "in the state of nature," or in "the savage state." Each man "appropriated whatever took his fancy and killed whomsoever opposed him, if he could." "Life was a continual free fight, and beyond the limited and temporary relations of the family, the Hobbesian war of each against all was the normal state of existence."[1]

It is hard to believe one's eyes when reading these words. Thomas Hobbes, forsooth! He was a philosopher who had published, two hundred years earlier, some sufficiently silly *a priori* anthropology. But Huxley is a great Darwinian scientist, and is writing in about 1890. Yet what he says is even sillier than anything that Hobbes dreamed up about the pre-history of our species.

What, for example, is a Hobbesian savage, presumably an adult male, doing with a *family* at all, however "limited and temporary"? In a "continual free fight," any man who had on his mind, not only his own survival, but that of a wife and child, would be no match for a man not so encumbered. Huxley's man, if he wanted to maximize his own chances of survival, and had even half a brain, would simply eat his wife and child before some other man did. They are first class protein, after all, and intraspecific Darwinian competition is principally competition for the means of subsistence, isn't it? Besides, wives and children are "easy meat," compared with most of the protein that goes around even at the best of times.

Huxley has even managed to burden Darwin with an absurdity which, though it was strongly suggested by Darwin's insistence on words like "struggle" and "battle," is by no means inherent in Darwinism itself. I mean, by his reference to "continual fighting." Fighting between conspecifics, even fighting over food, is not at all a necessary element in competition for survival as Darwin conceives it, whether it be humans, flies, cod or whatever that is in question. If you and I are competing for survival, and for ten days in a row you are able to get food while I cannot, then I starve to death and you win this competition, whatever may have been the difference between us which enabled you to win. Of course it *may* have been your greater fighting ability. But it might equally have been your superior speed, intelligence, eyesight, camouflage, or any one or more of a hundred other things. Fighting need never have come into the matter at all, as far as Darwinian theory is concerned. Which is just as well for that theory, since pines, most flies, and countless other species, *cannot* fight.

Huxley naturally realized that, as examples of Darwinian competition for life among humans, hypothetical ancient fights between Hobbesian bachelors were not nearly good enough. What was desperately needed were some *real* examples, drawn from contemporary or at least recent history. Nothing less would be sufficient to reconcile Darwinism with the obvious facts of human life. Accordingly, Huxley made several attempts to supply such an example. But the result in every case was merely embarrassing.

One attempt was as follows. Huxley draws attention to the fierce competition for colonies and markets which was going on, at the time he wrote, among the major Western nations. He says, in effect, "There! That's pretty Darwinian, you must admit."[2] The reader, for his part, scarcely knows where to look, and wonders, very excusably, what species of organism it can possibly be, of which Britain, France, and Germany are members.

A second attempt at a real and contemporary example was the following. Huxley says that there is, after all, still a *little* bit of Darwinian struggle for life in Britain around 1890. It exists among the poorest 5 percent of the nation. And the reason, he says (remembering his Darwin and Malthus), is that in those depths of British society, the pressure of population on food supply is still maximal.[3]

Yet Huxley knew perfectly well (and in other writings showed that he knew) that the denizens of "darkest England" were absorbed around 1890, not in a competition for life, but (whatever they may have thought) in a competition for early death through alcohol. Was *that* Darwinian? But even supposing he had been right, what a pitiable harvest of examples, to support a theory about the whole species *Homo sapiens*. Five percent of Britons around 1890, indeed! Such a "confirmation" is more likely to strengthen doubts about Darwinism than to weaken them.

A third attempt is this. Huxley implies that there have been "one or two short intervals" of the Darwinian "struggle for existence between man and man" in England in quite recent centuries: *for example, the civil war of the seventeenth century!* You probably think, and you certainly ought to think, that I am making this up; but I am not. He actually writes that, since "the reign of Elizabeth . . . , the struggle for existence between man and man has been so largely restrained among the great mass of the population (*except for one or two short intervals of civil war*), that it can have little, or no selective operation."[4]

You probably also think that the English civil war of the seventeenth century grew out of tensions between parliament and the court, dissent and the established church, republic and and the monarchy. Nothing of the sort, you see: it was a resumption of "the struggle for existence between man and man." Cromwell and King Charles were competing with each other, and each of them with everyone else too, *à la* Darwin and Malthus, for means of subsistence. So no doubt Cromwell, when he had had the king's head cut off, ate it. Uncooked, I shouldn't wonder, the beast. And probably selfishly refused to let his secretary John Milton have even one little nibble.

Huxley should not have needed Darwinism to tell him— since any intelligent child of about eight could have told him— that in a "continual free fight of each other against all" there would soon be no children, no women and hence, no men. In other words, that the human race could not possibly exist *now*, unless cooperation had *always* been stronger than competition, both between women and their children, and between men and the children and women whom they protect and provide for.

And why was it that Huxley himself swallowed, and expected the rest of us to swallow, this ocean of biological absurdity and historical illiteracy? Why, just because he could not imagine Darwinism's being false, while if it is true then a struggle for life *must* always be going on in every species. Indeed, the kind of examples for which Huxley searched would have to be as common as air among us, surrounding us everywhere at all times. But anyone who tries to point out such an example will find himself obliged to reenact T. H. Huxley's ludicrous performance.

There is (as I said earlier) a contradiction at the very heart of the Cave Man way out of Darwinism's dilemma: the contradiction between holding that Darwinism is true and admitting that it is not true of our species now. But I should perhaps emphasize that the absurdities which we have just witnessed in Huxley, though they no doubt were generated by that initial contradiction, are additional to it.

WHAT I CALL the Hard Man way of trying to reconcile Darwinism with human life is very different from the Cave Man way. The latter, as we saw, embraces the Darwinian theory, but then, in the case of man in historical times, illogically makes an exception to that theory. But the Hard Man despises that kind of feeble inconsistency: like the Earl of Strafford, his motto is "Thorough." He says that the Darwinian theory of evolution *is* true without exception, and it is just too bad for any appearances, that there are or may be in human life, which contradict that theory. They must be *delusive* appearances, that's all. Underneath the veneer of civilization, the Hard Man says, and even under the placid surface of everyday domesticity, human life is really just as constant and fierce a struggle for survival as is the life of every other species.

"Social Darwinists" is the name which is usually given to the people who take this way out of Darwinism's dilemma. But everyone agrees that it is a very inexpressive name. My name, "Hard Men," is preferable for several reasons. One is, that only a hard man, in the sense of a rigidly doctrinaire one, could possibly believe what these people do. Another reason is that *what* they believe implies that human life is an incomparably harder affair than anyone else has ever taken it to be.

Hard Men hold, then, that life among human beings is no less a ruthless competition for survival than it is among pines, or cod, or flies. It follows that in human life, as in the life of those other species, there is no care of the sick, the old, the poor, the afflicted, or the mad, and no protection of the innocent within a community, or protection of the community itself from hostile communities. In plain English: there aren't *really* any such things in human life as hospitals, charities either public or private, priesthoods, police, armies, or governments.

Of course no one, not even the Hardest Man, actually says

this, at least in print. You would have to be not just hard but mad to say so. All the same, it is what the Hard Man way out of Darwinism's dilemma really amounts to.

But what is it, then, that the Hard Men in their extensive writings do say? Why, this. Instead of saying, what according to their own theory, they should say, that unemployment relief (for example) is *impossible*, they say it is *deplorable*. (Because it actually increases poverty, both by rewarding economic dependence and by penalizing independence.) Instead of saying, what their own theory implies, that a hospital among human beings is *inconceivable*, like a hospital among flies, they say that hospitals are *injurious* to our species. (Because they enable unfit persons to survive and reproduce.) Instead of saying, what Darwinism really implies, that governments and priesthoods are *hallucinations*, they say that they are *harmful*. (Because they interfere with or negate the salutary processes of competition and natural selection.)

In this way a very curious historical fact has come about. Namely, that the writings of the Darwinian Hard Men make up, not at all what you would have expected, a literature of the biology and natural history of our species, but a literature of *moral and political exhortation* instead. Hard Men say that competition for survival, and the natural selection which results from it, are processes just as inevitable among humans as they are among pines or flies. Yet every page they write is written in order to prevent those processes being interfered with or negated: that is, *to prevent the inevitable being led astray!*

In fact the whole of Hard Man literature can be epitomized as follows.

People who are kind-hearted but ignorant of biology are always attempting, by means of such things as hospitals or unemployment relief, to suspend the law of the preferential survival of those organisms which are best fitted to succeed in the struggle for life. But they might as well try to suspend the law of gravitation, and the only result of their efforts, though also the invariable result, is a greater or lesser degeneration of the human stock.

In short, things like hospitals and unemployment relief are, at the same time, both impossible and injurious.

The inconsistency of this is not as immediately obvious as that of the Cave Man way out. Still, it is obvious enough. Or, in case it is not, I will say that the inevitable *cannot*—logically cannot—be led astray. If (for example) hospitals and unemployment relief really do interfere with or negate the processes of competition and natural selection, then those processes are not inevitable. If they *are* inevitable, then they really are inevitable, and there is not the smallest need for anyone to exert himself to prevent their being interfered with or negated. In particular there is no need for, and indeed no sense in, Hard Men writing books in order to warn us of the biological dangers of interfering with those processes. You *cannot* interfere with inevitable processes.

But of course this inconsistency has not stopped Darwinian Hard Men writing many books with that very purpose. One of the most influential of these, and one of the best too, is Herbert Spencer's *The Man versus the State* (1884).

This book is a powerful polemic against the encroachments of modern governments on the liberty of individuals. That is a real enough subject (to put it mildly) and one which is nowadays of rather more poignant interest than it was in 1884. But on every page of Spencer's book the characteristic Hard Man absurdity, of trying to prevent the inevitable from being led astray, lies like a tombstone. The evils which Spencer inveighs against are real, indeed. But they happen also to be ones which, if his own view of man were true, could not possibly exist.

Spencer's view of man is essentially Darwinian. But then, what is this thing, the state, doing in a Darwinian view of human life? How could there be a *state*, where there is constant, universal, unrestrained—and mostly unsuccessful—competition merely to *live*? Think of parallel cases. If a Darwinian writer, in giving an account of fly life, were to mention the existence of *fly hospitals*, everyone would see the absurdity at once. Similarly, if a Darwinian writer, in giving an account of pine life, were to tell us that there is a pine priesthood, or unemployment relief for "disadvantaged" pines. In the same way, there should be no mention, in a Darwinian account of human life, of such a thing as a state. From the Darwinian point of view, Spencer could just as sensibly have written a book called *The Fly versus the Fly State*, or *Pines Against Big Pine Government*.

And yet, for all its absurdity, it is very easy to understand how *The Man versus the State* came to be written. By 1884 the franchise in Britain had been extended to include virtually all adult males, with many results which could be easily foreseen. One of these results, which Spencer could see happening, and which by 1884 even a blind man could have seen, was that taxation was already obligating the middle and upper classes to have fewer children, in order that governments could support the irrepressible flood of offspring of the poor. But on the other hand, Darwinism says that population *always* presses on the supply of food, and that, from this pressure, competition for survival, and natural selection, must always ensue. Well, then, if Darwinism is an article of faith with you, as it was with Spencer, things like hospitals and unemployment relief are *bound* to look like wicked attempts to mislead the inevitable.

But it was the eugenics movement which was easily the most spectacular example of Darwinian Hard Men struggling manfully to keep the inevitable from going wrong. This movement stemmed originally from the writings of Darwin himself (although you were not then, and are not now, supposed to say so). But its official founder and leader, and the man who coined the word "eugenics," was his cousin, friend, and disciple, Francis Galton. By about 1880 Galton had become convinced, and had begun convincing others, that some eugenic measures—or what might be called measures of "quality control in humans"—were absolutely imperative for Britain.

The eugenists leave us in no doubt as to why they thought this. It was because, in late nineteenth-century Britain, the fittest people were visibly *not* outbreeding the less fit. In fact the boot was on the other foot. The overwhelming tide of philanthropic and egalitarian sentiment had brought about a population in which there was (as several writers put it at the time) a preferential "survival of the *un*fittest." That is, a preferential rate of reproduction by the indolent, the improvident, the unintelligent, the dishonest, the constitutionally weak, the carriers of hereditary disease, the racially inferior, and so on.

Of course other people might have drawn, from these same facts, a conclusion very different from the one that the eugenists drew. They might simply have concluded that Darwin's theory of evolution is false. After all, a eugenist does not have to be a Darwinian. Plato, for example, was a eugenist thousands of years before Darwinism was thought of.

Strictly speaking, there was one other conclusion which Darwinians could have drawn from the demographic facts which terrified them: namely, that the mentally defective, the carriers of hereditary disease, and so on, actually *were* fitter than the average upper middle class Britons. But this would have required the superior fitness of one group of organisms to another to be *identified with* its having a higher rate of actual reproduction: an idea which, though it is neo-Darwinian orthodoxy at the present day, really is as ridiculous as Galton would have thought it. For suppose it were true, and suppose Jack decides to have children by Jill, though he through genetic misfortune is blind, violent, and of sub-normal intelligence, while she has inherited deafness, syphilis, and AIDS. Then even the best medical advisor could only say to these intending parents something like the following: "It's no good asking me or anyone else whether you two are fit, or how fit you are. That can be known only after you have finished reproducing. If you manage to leave behind you more children than the average couple, that will prove you are fitter than the average couple, or rather it will *be* your superior fitness. But there's only one way to find out, so off you trot and get stuck into it. You could be lucky. Beethoven's father, remember, was a genetic disaster."

Galton and the other eugenists should really have concluded, then, from their demographic facts, that Darwinism is false. These facts (I should perhaps emphasize), though the eugenists certainly over-colored them, were real enough, and frightening enough too; just like that expansion of the state which very properly frightened Herbert Spencer. But of course it would have been entirely out of the question, a psychological impossibility, for someone like Galton to come to the conclusion that Darwinism is false.

Galton's intellectual and emotional situation was therefore this. On the one hand there was Darwin's theory of evolution. If it is true, then competition for survival is always going on in every species, and as a result natural selection is always going on too. Therefore, preferential survival of the organisms best fitted to succeed in the struggle for life *is* inevitable. But on the other hand there were, right before his eyes, the quite oppressive demographic realities of contemporary Britain. What could poor Galton possibly be expected to conclude, except that the inevitable *was* being led astray, and needed the help of people like himself in order to be put back on the rails?

Was Darwin himself free from this characteristic inconsistency of Darwinian Hard Men? Some justly respected writers imply that he was, and even that he was not really a Social Darwinist or Hard Man at all. But they are mistaken. Consider, for example, the following paragraph from *The Descent of Man, and Selection in Relation to Sex* (second edition, 1874).[5]

With savages, the weak in body or mind are soon eliminated; and those that survive commonly exhibit a vigorous state of health. We civilized men, on the other hand, do our utmost to check the process of elimination; we build asylums for the imbecile, the maimed and the sick; we institute poor-laws; and our medical men exert their utmost skill to save the life of every one to the last moment. There is reason to believe that vaccination has preserved thousands, who from a weak constitution would formerly have succumbed to small-pox. Thus the weak members of civilized societies propagate their kind. No one who has attended to the breeding of domestic animals will doubt that this must be highly injurious to the race of man. It is surprising how soon a want of care, or care wrongly directed, leads to the degeneration of a domestic race; but excepting in the case of man himself, hardly any one is so ignorant as to allow his worst animals to breed.

This will be admitted to be the utterance of a Hard-*enough* Man, at any rate. In particular, it is plainly the utterance of a eugenist. Yet it was published in 1874: that is, at a time when eugenics was hardly even a gleam in Francis Galton's eye. Nor is it the utterance of one of the softer eugenists, either: think about Darwin's reference here to the singular folly of "*allowing one's worst animals to breed.*"

But if further evidence is needed that Darwin was, sometimes at least, a Darwinian Hard Man, the following two paragraphs will supply it.

Man scans with scrupulous care the character and pedigree of his horses, cattle, and dogs before he matches them: but when he comes to his own marriage he rarely, or never, takes any such care. He is impelled by nearly the same motives as the lower animals, when they are left to their own free choice, though he is in so far superior to them that he highly values mental charms and virtues. On the other hand he is strongly attracted by mere wealth or rank. Yet he might by selection do something not only for the bodily constitution and frame of his offspring, but for their intellectual and moral qualities. Both sexes ought to refrain from marriage if they are in any marked degree inferior in body or mind; but such hopes are Utopian and will never be even partially realized until the laws of inheritance are thoroughly known. Everyone does good service, who aids towards this end. When the principles of breeding and inheritance are better understood, we shall not hear ignorant members of our legislature rejecting with scorn a plan for ascertaining whether or not consanguineous marriages are injurious to man.

The advancement of the welfare of mankind is a most intricate problem: all ought to refrain from marriage who cannot avoid abject poverty for their children; for poverty is not only a great evil, but tends to its own increase by leading to recklessness in marriage. On the other hand, Mr. Galton had remarked, if the prudent avoid marriage, whilst the reckless marry, the inferior members tend to supplant the better members of society. Man, like every other animal, has no doubt advanced to his present high condition through a struggle for existence consequent on his rapid multiplication; and if he is to advance still higher, it is to be feared that he must remain subject to severe struggle. Otherwise he would sink into indolence, and the more fitted men would not be more successful in the battle of life than the less gifted. Hence our natural rate of increase, though leading to many and obvious evils, must not be greatly diminished by any means. There should be open competition for all men; and the most able should not be prevented by laws or customs from succeeding best and rearing the largest number of offspring.[6]

Of course it would be easy to find, in other authors, *Harder*-Man utterances than these. For example, in some of the writers who are quoted in R. Hofstadter's *Social Darwinism in American Thought* (1959): writers, that is, who used Darwinism to justify the economic activities of the "robber barons" of American capitalism about a hundred years ago. Not that even any of *them* was the Hardest Man of all Hard Men. That distinction belongs, as far as I know, to Adolf Hitler, an instructive quotation from whom can be found in M. Midgley's *Evolution as a Religion* (1985).[7]

Some other Hard Men, then, would undoubtedly go further than Darwin does in the passages just quoted, and would add other things which he would not at all have agreed with: perhaps some idiotic proposition about the racial inferiority of Jews, for example. But is there anything in the above passages which even the most rabid of the American Social Darwinists, or even Hitler himself, would have *disagreed* with? If there is, I have failed to detect it. These passages are, in fact, just standard issue Hard Man material, and contain its standard issue inconsistency. For they combine, in about equal proportions, suggestions that man is inevitably subject to natural selection, and suggestions that we will have to be right on our toes to make sure he stays that way.

The last two paragraphs quoted reveal, in addition to Darwin's eugenism, his opposition to the practice of contraception. Nor was he, as one might have expected, opposed to contraception only where what he calls "the better members of society" were concerned. He was emphatically opposed to contraception altogether. He wrote, in reply to a correspondent who had expected him to have a very different attitude to that subject, that "over-multiplication [is] useful, since it cause[s] a struggle for existence in which only the strongest and ablest survive. . . ."[8] So it is clear enough that Darwin considered contraception to be one of the dangers threatening to overwhelm the inevitable.

A Hard Man nowadays (like a good one according to the old song) is hard to find, at least in print: certainly far harder than a hundred years ago. But unfortunately it is impossible to determine how far this fact is due to a real change in what people believe. It may be more due to a change in what they are allowed to say. Our freedom of the press, except for really precious things like pornography, has greatly diminished in the last hundred years, and especially in the last twenty. In 1892 you could say in public print that women are intellectually inferior to men, that blacks are morally inferior to white, that poor people are lazier than middle class ones, that Shi'ite Moslems are ignorant murderous fanatics, and so on. You cannot say so now. Or if you do by some fluke manage to get something of that sort into print, you will need to revise your own and your family's insurance policies, the terms of your employment contract, and your home security.

So if nowadays Darwinian Hard Men are seldom to be met with, or at least to be identified, in print, the reason may simply be that they, and their editors and publishers, are frightened of such powerful and ruthless groups as feminists, blacks, Shi'ites, etc. My opinion, for whatever that is worth, is that this *is* the main reason for the apparent scarcity of Hard Men. In reality, I suspect, there are still plenty of them, especially among those neo-Darwinians who have come to be called "sociobiologists." They are too scared to say what they think, that's all.

WHAT I CALL the Soft Man way out can be quickly dealt with. Strictly speaking, it is not so much an attempt to resolve the inconsistency between Darwinism and human life, as a mere failure to notice that there is any inconsistency to be resolved.

The Soft Man is intellectually at ease. Having been to college, he believes all the right things: that Darwin was basically right, that Darwin bridged the gap between man and animals, etc., etc. He also believes, since he is not a lunatic, that there are such things as hospitals, welfare programs, priesthoods, and so on. But the mutual inconsistency of these two sets of beliefs never bothers him, or even occurs to him. He does not think that his Darwinism imposes *any* unpleasant intellectual demands on him. So he is not drawn to postulate, for example, as a concession to Darwinism, a period even in the remote past of all-out competition among people. He leaves that kind of thing to some of the television cartoons that five-year-olds watch. Still less does he think that his Darwinism requires him to advocate eugenics, or to oppose welfare programs, as the Hard Men do. In fact the politics of Darwinian Hard Men fill Soft Man with horror. They do, at any rate, until the suburb where he lives is taken over by blacks, or Shi'ite Moslems, or Croats, or Sikhs, or whatever.

The Soft Man is certainly the most appealing of the three ways out of Darwinism's dilemma, if we all agree to call it such a way at all. Utter helplessness almost always has something very appealing about it, and intellectual helplessness is no exception to this rule; while Soft Man is an extreme instance of such helplessness, or (in Samuel Johnson's phrase) of "unresisting imbecility."

But then, I do not really need to introduce Soft Man to you: you know him well. And the reason is, that he is you, and you, and—most of the time—me. We freeze to the marrow when we remember the hardest of all Darwinian Hard Men, and his gas ovens. But we also think that the person who put us basically right about man was the one who wrote, in a discussion of human life, of the unparalleled folly of allowing one's worst animals to breed.

I may add that Soft Man was also Charles Darwin himself: a fact which many Soft Men regard as a quite good enough excuse for the chaotic state of their own opinions about human life. Darwin's personal recipe for resolving Darwinism's dilemma was a mixture, in roughly equal proportions, of Hard Man and Soft Man, with just a dash—say 10 percent—of Cave Man thrown in. At the present day the most admired Darwinian chefs prefer to go a bit easier on the Hard Man ingredients; though that may simply be due, as I have suggested, to fear. Anyway, these disagreements among the experts are not so great that they need to concern mere street-Darwinians like ourselves.

Essay 2
Where Darwin
First Went Wrong About Man

[E]very single organic being around us may be said to be striving to the utmost to increase in numbers.
—Charles Darwin, *The Origin of Species*

I

EASILY THE MOST CELEBRATED DATE in the history of the theory of evolution is 1859, because it was in that year that Darwin published *The Origin of Species*. That event fully deserves the celebrity which has been bestowed on it, and hence on the year 1859. But a question which cries out for an answer is this: why was it left as late as that for some such book as Darwin's to appear?

By 1859 the fact of evolution—the fact that new species arise (when they do) out of old ones—had been staring naturalists in the face for decades. Even by about 1835, there was simply no other natural interpretation of the fossil record. And even as regards our own species, it was plain enough by 1835, from embryology, and from comparative anatomy and physiology, that we must be connected by descent with other kinds of animals.

The idea of evolution (as is by now well known) had been more or less "in the air" for about eighty years by the time

Darwin published his book. People did not then *call* it "evolution"—they called it "development" or "descent with modification"—but the idea was certainly the same. And this idea, with every other passing decade after about 1815, came to haunt the minds of naturalists more and more. Yet even so, believers in evolution continued to be only a tiny minority, even among naturalists. The most confident and explicit evolutionists, unfortunately for their own cause, were not naturalists at all: Robert Chambers, for example, the then-anonymous author of *Vestiges of the Natural History of Creation* (1844). Of the even tinier minority who were naturalists, some, such as Buffon, made a habit of asserting the reality of evolution, and then later denying it. Others, such as Lamarck and William Lawrence, had asserted the reality of evolution unequivocally enough, but had unluckily made it part of a "package deal" with other and less defensible ideas. Yet others again, such as Erasmus Darwin (Charles' grandfather) had in effect made a large *exception* to their evolutionism, by their deafening silence on the delicate subject of *human* evolution.

So as things turned out, although biological knowledge had been converging irresistibly on the same conclusion for decades beforehand, it was in fact left to Charles Darwin to say, in 1859, clearly and consistently and without the introduction of any extraneous matter, that all existing species have evolved from earlier ones. He expressly included man in this generalization. But at the same time—it should be remembered—he also took care to say, in *The Origin of Species*, not one word more on the subject of that interesting species.

There were three things which made Darwin's fellow naturalists reluctant to admit the fact of evolution. One of them is well known: it was religion. The Book of Genesis says that the organisms we see around us had all been created, just as they are now, by God. That cannot be true if in fact they have all developed out of older species. It is quite wrong, however, to think of the religious objection to evolution simply as a matter of timidity on the part of the naturalists, or of repressiveness on the part of Church authorities. In the year 1835, for example, naturalists as a class were not any less religious than educated people in general. Most of them were understandably reluctant, therefore, to say or even to think that the book which they regarded as the Word of God was false.

A second thing which made Darwin's fellow-naturalists reluctant to admit the fact of evolution is one which, unlike religion, has been almost entirely forgotten. It was a *moral* objection, and a well-founded one. The idea of evolution was a brain child, and a representative one, of the French Enlightenment of the last quarter of the eighteenth century. In the minds of most naturalists in 1835, therefore, evolutionism was inextricably associated, and rightly associated too, with revolutionary republicanism, regicide, anti-religious terrorism, and the deliberate destruction, for the sake of equality, both of thousands of innocent people and of high culture in any form. A revolutionary judge, as he sent Lavoisier to the guillotine in 1794, said "The Republic has no need of chemists." Nor did the evolutionism of his late father suffice to save the son of Buffon from the same fate in the same year. But then, the Buffons were aristocrats, and by 1794 Robespierre had decided, and announced, that atheism is a distinctively *aristocratic* vice.

These being the circumstances, the reluctance of most naturalists in the first half of the nineteenth century to admit the fact of evolution was not only understandable: it was morally to their credit. It was not creditable to their heads; but to their hearts, it was. Consider, by way of contrast, that dedicated evolutionist and complete child of the Enlightenment, Erasmus Darwin. Though he lived until 1802, he had never wavered for one moment in his admiration for the French Revolution, or doubted that it was a guiding light for other nations to follow. He never suffered a single qualm, however much strange fruit the guillotine tree might bear. By comparison with this man, the great majority of British naturalists, who were all Christians and anti-evolutionists, have left a far cleaner smell behind them.

When Charles Darwin was born in 1809, therefore, evolutionism still stank of the Terror of 1793. Ever since 1789, of course, there had been in Britain an active minority of Enlightened persons, such as his grandfather, who were anxious to import to their own country all the blessings, including evolutionism, of revolutionary France. These people suffered a severe depression of their hopes, naturally, in the twenty years of intermittent war with France, between 1795 and 1815. But then, at Waterloo, all hopes of France's exporting Enlightenment by force of arms were extinguished. And with this, the old package deal, of evolutionism with anti-religious, republican, and democratic fervor, at once sprang to life again. In fact, it proceeded to flourish as never before, and threw up new manifestations everywhere with irresistible exuberance. By about 1830 it would have been as easy to find an evolutionist who was a loyal member of the Church and subject of the Crown, as it would be to find a "green" ideologue today who is a bulldozer enthusiast, or to find an Orthodox pig farmer in Israel.

Darwin, consequently, when he became convinced of the reality of evolution in the late 1830s and the early '40s, found himself faced with a task of some delicacy. In order to tell the public what he knew, and yet not incur extreme and deserved odium, he needed to separate evolutionism from the swarm of murderous associates which up to that time had always accompanied it. He succeeded in doing so too, though only by the exceedingly drastic method of saying, in *The Origin of Species*, nothing whatever about the origin of the most interesting species of all: man. No doubt he found this improbable silence the easier to maintain, because his own temperament was pacific, and because his moral and political ideas were not at all utopian, but just moderate Whig. Anyway, nothing could exceed the circumspection with which he went about the task of separating evolutionism from its original matrix of irreligion and revolution. Even after the huge success of *The Origin*, he let twelve more years pass before he first ventured to handle the subject of man in print (in *The Descent of Man*, 1871).

The third reason why most naturalists around 1835 were slow to

admit the fact of evolution was neither a religious nor a moral objection. It was a purely intellectual one. By now it has been almost
completely forgotten, no doubt because we labor under the hand-
icap of hindsight. But it was a well-founded objection at the time.

If someone says a certain thing has happened, and it is of a
kind which has never been actually witnessed by anyone, it is
reasonable to doubt what he says, if no one can think of any *explanation* of what he says has happened. It is on this principle
that you would doubt my word, if I were to tell you (for example) that electrical storms follow me wherever I go. Now this
was exactly how matters stood with evolutionism around 1835.

No naturalist claimed, of course, to have ever *seen* a new
species evolve out of an older one. Yet the evolutionists said that,
whenever new species do come into existence, that is the way
they do it. But what could be the *explanation* of one species'
giving rise to another? What causes or forces are there, already
known to exist in nature, which would make one kind of grass or
fish or mammal evolve into a different kind? Where is the *vera
causa*, as they used to say, or (as we would say), where is the
mechanism, which could *drive* this alleged process of evolution?

It should go without saying that this was not only a purely intellectual objection, but a good objection, to evolutionism. The
main evidence *for* evolution was the fossil record, which reveals in
countless instances the arrival of a new species which is closely related to an earlier one. In 1835 most naturalists regarded these
new species as brought about by exercises of God's creative power;
whereas the evolutionists regarded them as developments or
evolutions of the older species in question. No one had ever *witnessed* any of these exercises of Divine power, of course, but then
exactly the same was true of evolutions of one species into
another; no one had ever witnessed an instance of that, either. And
then, to ascribe new species to God's creative power is at least an
explanation of a kind, though doubtless not of a very satisfactory
kind. But the evolutionists, for their part, had *no explanation of
any kind* to suggest for their alleged process of evolution.

Darwin, being a rational man, naturally felt the force of this objection, just as strongly as did his fellow naturalists who were not evolutionists. For several years around 1836, it weighed heavily on his mind. These were the very same years when the *reality* of evolution was being constantly impressed upon him, by the multitude of facts which would be explained if it were true. But the trouble, and a very big trouble, was that he could not think of anything which would *explain evolution. That* was the rub, and it seemed to Darwin that he was staring at a blank wall.

Given the intellectual circumstances of the time, it is not surprising that, just a few years later, another young naturalist found himself brought to a standstill by exactly the same blank wall. This was Alfred Wallace. Though neither of them knew it, his early intellectual career had been exactly the same as Darwin's. On the one hand, he had become convinced of the reality of evolution; but on the other, he was altogether at a loss as to how to explain it.

Why should there be any evolution at all? Why should not the species which exist at a given time exist forever, without any new ones ever being added, or old ones subtracted? But it is not the subtractions which are the problem: presumably climatic or topographical changes, and general wear and tear, will sometimes bring about the extinction of a species. The problem is the new additions. Why should any new species ever come into existence at all? That is the mystery of the origin of species, which both Darwin and Wallace long brooded over in vain.

To ordinary observation, of course, it does not look as though new species ever do come into existence. But it is clear from the fossil record that the reality is very different. In countless thousands of instances, new species of organisms have appeared on earth. Organic nature is in fact, whatever else it is, a gigantic *species-generating* engine. Now, why in the world would it be that? What force can it possibly be, which drives this gigantic engine? It might reasonably be thought to be some Divine force, in view of the irresistibility of its operations, and the length of time that those operations have been going on all over the earth. But if it is not a Divine force, what force is it?

II

Darwin found the answer to this question, or at least the answer which satisfied him, in a most unexpected place. Namely, in *An Essay on the Principle of Population* by the Reverend T. R. Malthus, which had appeared first in 1798. By an extraordinary coincidence Wallace too, a few years later, independently found the same answer by reading the same book; though he was much slower than Darwin had been to realize that Malthus *had* supplied the answer.

Nothing could have been further from Malthus's mind, when he wrote his *Essay on Population*, than explaining why there is evolution. In fact, he would have been appalled if he had lived long enough to learn that he had, through Darwin and Wallace, opened the way for the triumph of evolutionism. But he died in 1834. By that time his 1798 *Essay*, much expanded, had gone through five further editions, and had exercised great influence. But it did not contain one word about evolution, and was not intended as any sort of contribution to biology. It was intended to be, and was, an economic and political tract for the times.

The *Essay on Population* was a counter-blast to all the Enlightened visions of the future which had been pouring out of France for fifty years by the time that Malthus wrote: visions of the universal happiness, equality, communism, sexual emancipation, etc., which were going to ensue once religion, monarchy, and private property had been overthrown. By 1798 there were many people not only in France but in Britain (including Malthus's own father) who were completely under the spell of these utopian ideas. But Malthus, like most decent and intelligent Englishmen of his time, could see that it was those ideas which had brought about the French Revolution, the Terror, and the desolation of Europe by Enlightened French armies. He wished to save Britain, and Europe, from the fate which had overtaken France. And he believed he had detected, in all of these optimistic visions, a fatal flaw which had previously gone unnoticed. It was in order to point out this flaw that he wrote his *Essay*.

Nor, did Darwin, for his part, open his book with any idea that it might enable him to explain evolution, or even that it might help him in any way at all with his biological inquiries. In fact he read it to take his mind off those things. He says in his *Autobiography* that "in October 1838 . . . I happened to read for amusement 'Malthus on Population.' . . ."[1] What led Wallace to read Malthus in about 1846 I do not know; but since it took more than another ten years for the significance of the *Essay on Population* to come home to him, he certainly could not have had *in advance* any inkling of what he was ultimately to find there. Anyway, both Darwin and Wallace did find in Malthus, to their great joy, the key to the explanation of evolution. Nor did either man, ever afterwards, waver in his belief that he *had* there found that key. Darwin had a lifelong bad habit of not acknowledging, until he was obliged to do so, the debts his work owed to other people; either that, or he had a still worse habit of not even noting them. But his debt to Malthus was so great that even Darwin could not have failed to notice; and Malthus's book was so widely read that thousands of people would at once have detected Darwin's indebtedness to it, if he himself had failed to notice it. In fact, he did acknowledge the debt from the very first edition of *The Origin of Species*, not only in its third chapter, but in the introduction.

The key in question was the proposition that, in every species of organisms, population always presses upon the supply of food available, and tends to increase beyond it. According to Malthus, a population of organisms, whether they are humans or cod or pines or whatever, is always as large as its food supply allows it to be, or else is rapidly approaching that limit. It makes no difference whether the population is large or small, dense or sparse, or whether it is increasing, decreasing, or stationary. In all species, the tendency to increase in numbers by reproduction is so strong and constant that, whenever there is food for a possible pine, cod, or human, there is, or else soon will be, an *actual* pine, cod, or human.

This proposition is what Malthus meant when he spoke of "the principle of population." It had first been suggested by observations of the prodigious number of viable seeds which are produced each year by a single adult pine tree, of eggs which are produced each year by a single adult female cod, and so on. Some simple calculations, based on these observations, had revealed that astonishingly few years would be enough for cod to fill the oceans, pines to cover all soil, etc., if every *potential* parent cod survived to become an actual one, or if the vast majority of viable pine seeds were not prevented, by early death, from realizing their full reproductive potential. These observations and calculations, which appear to have originated about fifty years earlier, were not alluded to by Malthus in his first edition of 1798. But in the very different second edition of 1803, and in all the subsequent editions, they *were* alluded to, and put where they belong, too: in the first two pages of the book.

It was almost certainly in one of these later editions that both Darwin and Wallace read Malthus. Darwin's own copy of the *Essay on Population* was of the sixth edition, of 1826. By the 1830s and '40s, copies of the 1798 edition must have been much scarcer than copies of one of the five later editions. By that time, in addition, the 1798 book would have seemed like a period piece, because of its preoccupation with such forgotten visionaries of the 1790s as Godwin and Condorcet. But whatever the details of the literary transmission may have been, it is certain that, *via* Malthus's *Essay*, some of those observations and calculations, about the astonishing number of various organisms which would exist if their natural tendency to increase were unchecked, found their way in 1859 into the vital Chapter III of *The Origin of Species*.

From there, they have gone on to become part of the mental furniture of every educated person. Nor have they, even now, lost their power to astonish us. Among organisms in general, the strength of the tendency to increase by reproduction, and the strength of the resulting pressure of their numbers on their food supply—even, if you like to call it so, the strength of the Life Force—*is* astonishing. There is nothing metaphysical or mystical about this; and nothing merely verbal either. *A priori*, zero population growth might have been a universal law of organic nature. But the fact, as it happens, is very far otherwise.

Now, variation is universal among organisms. Every organism is always different in some respect from every other, even from those most closely related to it. Some of these variations are transmissible to offspring. And some of them are such as to bestow a certain advantage on the organisms that possess them: superior strength, better vision, greater speed, or whatever it might be.

These facts about variation are all very obvious, and must have been noticed countless times. In themselves they are quite uninteresting facts. But the moment they are combined with Malthus's principle of population, they leap into explanatory life.

If population constantly presses on the food available, every new generation of organisms must always find (as Malthus said)[2] that the places at the table of life are already full, or nearly full. There will therefore be competition, among the members of each generation, to occupy a place: literally a struggle for life among them. The competitors being so numerous, and the vacant places so few, most of the competitors *must* fail. Nor are they equally well-equipped for success in the competition into which they find themselves born. Some of them will possess, by one of the myriad accidents of variation, some advantage that others lack. These ones will, on the average, succeed better in getting food, and leave more descendants, than their less fortunately endowed fellows do. Thus will emerge, within a species, a "favored race" or variety, distinct from the ancestral type, and reproducing at a faster rate. This race will therefore have, even before it reaches equality in numbers with the species type, a better chance of benefiting from *additional* advantageous variations, as these crop up in a random way across the species. "To him that hath, shall be given." As advantages accumulate in this race or variety, it will become more and more distinct from the type. When it has become, in virtue of its accumulated advantages, so distinct from the type that the two cannot any longer breed together, the favored race will have become, in fact, a new species.

This was the explanation of evolution, or the origin of new species—"by means of Natural Selection, or the Preservation of Favored Races in the Struggle for Life," as Darwin put it— which came like a revelation over the minds of Darwin and Wallace, once they had each absorbed Malthus's *On Population*. Even at this distance of time, it is hard not to share some of the exhilaration which they undoubtedly felt, as this reasoning unfolded itself before their minds. It is like all the best reasoning: *natural, without being obvious*. Huxley, after he first read *The Origin of Species*, exclaimed, "How extremely stupid not to have thought of that!" His annoyance was understandable; but it was not just. No doubt Pythagoras's disciples, after their master had first proved the famous geometrical theorem which still bears his name, felt just as Huxley did. But Pythagoras's reasoning, like that of Darwin and Wallace, is *not* obvious to ordinary minds: just natural, once they have been shown it.

Then, how admirably *prosaic* this explanation of evolution is! It must have seemed to Darwin and Wallace that they had wakened to "the light of common day," after a long night of obscurely frightening dreams about Divine energies. The question had been, what forces or causes can it be, which makes existing species generate new ones? And the answer turns out to be, forces no more awe-inspiring than the tendency of organisms to multiply, and the necessity for them to get food in order to survive: two sufficiently familiar things!

The only element which remains truly mysterious, in the early history of the theory of evolution, is this. Why did no one before the mid-eighteenth century ever realize the tremendous strength of the tendency of organisms to increase in numbers, or of the resulting pressure of their population upon their food supply? Malthus (as I have implied) was certainly not the first person to realize these things. There are anticipations of his "principle of population" in the writings of David Hume, Benjamin Franklin, Joseph Townsend, and no doubt others beside; but not, or not to any extent worth mentioning, in any writings whatever before about 1750. And yet people could have made, at any time during thousands of years before that date, at least a rough comparison between the size of a batch of fertilized cod eggs, or viable pine seeds, and the number of this batch which survived to reproduce in turn.

But whatever may have been the reason for it, it *was* left to Malthus to teach naturalists the strength of the organic tendency to increase, and of the resulting pressure of their numbers on their food. And he happened to do so in a book which, for reasons quite unconnected with evolution, reached an unusually great number of readers. He thus unintentionally provided Darwin and Wallace with their explanation of evolution, and hence, indirectly, with the key to all the lower level explanatory successes which their theory went on to enjoy. There was no more fertile idea in all biology, before the present century, than that "principle of population" which Darwin got from what he rightly called Malthus's "ever-memorable *Essay*."[3]

There was a cruel irony in this affair. For Malthus was, along with Edmund Burke and Joseph de Maistre, one of the bitterest enemies, and wisest critics, of the Enlightenment; while evolutionism (as I have said) was a regular element of the Enlightenment's intellectual armory. Yet in the 1830s and '40s when evolutionists had got hopelessly stalled by the problem of explaining evolution, it was Malthus, and he alone, who provided them with the explanation which they had been seeking in vain. Once it was fitted with the vital part that Malthus supplied, the evolutionary locomotive sped away on its headlong and triumphant career, as it has continued to do to the present day.

III

Darwin's explanation of evolution, then, and Wallace's, was as follows. "In every population of organisms, there is always variation, some of which is heritable and advantageous to its possessors, and there is always pressure of population on the supply of food, which results in a constant struggle for life among conspecifics. In this struggle, those organisms which possess some heritable advantage over their rivals will be 'naturally selected,' and in time, from being a favored variety of an old species, will become a new species."

In 1859, this was the best explanation of evolution available, and hence, indirectly, the best available explanation of the many facts which evolution in turn explains: the adaptation of organisms, their distribution, their affiliations with other species existing or extinct, and so on. *It is still the best explanation available of all those things.*

That is under-praising it, however, because the best available explanation of something need not be a good one. But the Darwinian explanation of evolution is a very good one as far as it goes, and it has turned out to go an extremely long way. Its explanatory power, even in 1859, was visibly very great, but it has turned out to be far greater than anyone then could have realized. And then, in the 1930s, the Darwinian theory received further accessions of explanatory strength through its confluence or synthesis with the new knowledge of genetics. And this "new synthesis," or "neo-Darwinism," has been itself growing rapidly in explanatory power ever since.

Still, it is obvious that the best available explanation of certain matters might yet be false or incomplete, even if it is a very good explanation as far as it goes, and it goes a very long way. It might be only the closest approach that we have yet made to a true and complete examination of the matters it is intended to explain. Moreover, even the best available explanation need not be equally good at all points. For *some* of the matters it is meant to explain, a certain theory might be a good approximation, or even be the complete and exact truth, while being at the same time glaringly incomplete, or even obviously false, with respect to some of the other things it is meant to explain.

That is, I believe, the way matters *actually* stand with neo-Darwinism. In particular, I believe that neo-Darwinism, though a very good approximation to truth and completeness for many of the simplest organisms, is an extremely poor approximation in the case of our own species. Or rather, to tell the truth, I think that it is, at least in the hands of some of its most confident and influential advocates, a ridiculous slander on human beings. I hope to convince readers of this in some of the later essays in this book. In the present one, I hope to convince them that the trouble began much earlier: namely in 1838, when Darwin embraced Malthus's principle of population.

THE DARWINIAN EXPLANATION of evolution, as we have seen, rests on two propositions. One of them is about variation. The other proposition is that any population of organisms is always pressing upon, or tending to multiply beyond, its supply of food; in other words, that every organic population is always as large as the available food permits, or else is rapidly approaching its limit.

I will call this "the Malthus-Darwin principle." As long as it is understood to be subject to an "other things being equal" proviso, and not mistaken for an exact and categorical truth, this principle is (as I have implied) a proposition of immense explanatory value, and one which was inexplicably overlooked until the second half of the eighteenth century. The principle's explanatory value is the greater, the more primitive the species of organism in question. But it has considerable explanatory power even in relation to man, and even in relation to man's history as distinct from natural history. For example, the speed with which human populations typically return to their former size, after even the most devastating wars or plagues, always amazes most people. But people who are apprised of the Malthus-Darwin principle are seldom surprised by it.

Taken just as it stands, however, and not subject to any proviso, the Malthus-Darwin principle is false. Indeed, it is obviously false, a thousand times over. Everyone knows of many organic populations which by no means obey this principle.

Domestic pets, for a start. Consider, for example, the population of cats, or goldfish, or dogs, which are living at this moment in New York apartments. Almost all the members of these populations are well fed. But millions of them, probably the majority of them, never reproduce at all. Hardly any, it is safe to say, have as many offspring as the available food would support.

As a counter-example to the Malthus-Darwin principle, this is sure to be considered not only trivial and unfair, but lacking in dignity. In the stark and unfeeling studies of evolutionary biologists, it will be felt, there is no place for such emotion-laden frivolities as domestic pets. But this is ridiculous, and just too bad for those studies. The population of cats in present-day New York apartments is just as respectable a biological population as any other. It is even perfectly possible that something of value to evolutionary biology might be contributed by a New York veterinarian who specialized in (say) the epidemiology of urban felines. Fortunately, however, we do not need to decide whether domestic pets are, or are not, entitled to vote against the Malthus-Darwin principle. For they are in any case only the tip of the anti-Malthusian iceberg.

Another bit of the iceberg is equally well known. This consists of the many species of animals and birds that are not at all domestic pets—some of them as far as possible from being so—which even though abundantly supplied with food, reproduce sparingly or not at all in captivity. If the Malthus-Darwin principle were true, there would not be the big money there is in illegally exported Australian birds. But, as overseas bird-fanciers know to their sorrow, that principle is not true. So those fanciers, when their old birds die, have no alternative but to buy a new pair.

Then, the population in the huge African wild animal reserves are neither domestic pets nor in captivity, but they sometimes fail to increase in numbers, or even decline, in the presence of abundant food. Such cases call for special explanation, of course; but then I am not denying that the Malthus-Darwin principle is a valuable explanatory rule, *other things being equal*.

Further non-Malthusian populations are all the countless selected flocks or herds or plots—of sheep, cattle, wheat, or whatever—which are maintained by breeders for commercial purposes. The merinos in a prize Australian flock, for example, are not domestic pets, are not wild animals in a reserve, and yet are not in captivity either: there are no such things as feral sheep. These merinos are, of course, well fed, but they by no means increase up to the limit that the available food would allow. For the breeders rigorously cull the offspring with an eye to maintaining or improving the quality of the flock.

Other populations which are well fed but do not obey the Malthus-Darwin principle are most of the experimental animals or plants which scientists maintain, not for profit but in order to learn something new in plant genetics, or whatever it might be.

Clinical microbiologists, thousands of times every day, "culture" populations of pathogenic bacteria for medical purposes. These populations are for a time well fed, and during that time they usually increase in numbers just as the Malthus-Darwin principle leads one to expect. But then—normally, and barring accidents—their existence is terminated, but not by their being starved to death. When extinction overtakes them, they may well be still swimming in a nourishing medium.

There are, then, countless populations of organisms which violate the Malthus-Darwin principle. But it will be objected, or course, that all the cases I have mentioned are vitiated by involving some kind of human intervention. The Malthus-Darwin principle (it will be said) cannot be refuted by instances of that kind, because it is a generalization only about organic populations in their *natural* state.

When a population in one of the African reserves fails to obey the Malthus-Darwin principle, there is not, in general, any reason to believe that human influence *does* have anything to do with the matter. Still, all my other counter-examples do contain the element of human influence; so let us for simplicity suppose that all of them without exception do. That is, that whenever there is a non-Malthusian relation between population and food, some kind of human influence if at least partly responsible.

But the awkward question is, how does the presence of human influence prevent these cases from being *natural* ones? This question is especially awkward, I may observe, for Darwinians. Man is one species of animal among others. If there is anything which is natural *to man*, it is having domestic pets, keeping animals in captivity, maintaining select populations of animals or plants for economic or intellectual profit, and cultivating pathogenic bacteria for the purpose of diagnosing and treating disease. These are simply some of the innumerable transactions which take place between members of our species and others, such as cats, sheep, wheat, or bacteria. But how can one and the same transaction, between our species and another, be natural at the man-end of it, and yet not natural at the other? This is a question which anyone will have to answer, if he hopes to rule out my counter-examples as not natural. But he will find it is a question more easily asked than answered.

It is well known that disagreements about what is natural and what is not are peculiarly likely to be unsettlable. In an attempt to detour such disagreements, we might be tempted to qualify the Malthus-Darwin principle, by saying that every population is always as large as its food permits, or is rapidly approaching that limit, *except where human influence prevents that being the case*. But this is just as obviously false, or rather ridiculously false, as the unqualified principle.

As everyone knows, non-human species are constantly engaged in restricting or preventing the increase of other species, by means other than limiting the food which is available to them. They do so by predation, for one thing, and by parasitism, for another. The dodo was surrounded by food when it died out, not from hunger but from predation. It was, as it happened, from human predation, but the extinction could very easily have been brought about by some other predator, or by a parasite. Flightless birds are "sitting ducks" both for predators and for soil-borne parasites. Likewise, it is reasonable to suppose, many of the countless extinctions which occurred before man existed took place in the midst of plentiful food.

It is only too likely that the reader will be inclined to infer, from the recital of commonplace facts which are inconsistent with the Malthus-Darwin principle, either that Malthus and

Darwin were irrational in believing that principle, or (more likely) that I am mistaken in ascribing the principle to them. Both of these inferences would be mistaken. But to *show* that they are mistaken would make the present section of this essay disproportionately long. I have therefore postponed my attempt to do so to Essay 3 below.

AS WELL AS straight counter-examples to the Malthus-Darwin principle, there are other objections to it which are only a little less direct and obvious. One of these concerns the commonness, or otherwise, of incestuous reproduction.

If a population is to be always as numerous as its food supply allows, or nearly so, reproduction would always have to begin as early as possible. In nearly all species of animals, all the earliest opportunities for mating are opportunities for the young to mate with a sibling or with one of their parents. You would expect, therefore, if the Malthus-Darwin principle were true, to find throughout the animal world a distinct bias towards incestuous reproduction, at least during early adulthood.

In fact, however, there is not only no such thing: there is the very opposite—a marked and general bias *against* incestuous reproduction. That this bias is not strictly universal should go without saying. But it is so very general and strong that biologists, ever since Darwin, have believed that they could even see a prefiguring or parallel of it, in the great trouble plants go to, to prevent their flowers from self-fertilizing. Whether or nor they are right in this, it is simply a fact about most animals that on the whole incestuous reproduction, which ought to be "the most preferred option" if the Malthus-Darwin principle were true, is the least preferred.

The same objection can be generalized. If a population is always at or near the maximum size that its food allows, then neglected opportunities for reproduction must always be at or near their minimum number. But again, the contrary is the case. Animal life in fact swarms with neglected opportunities for reproduction. The unmated adult female birds who act as "aunts" to the offspring of others are one well-known example. The young males of many species whose reproduction is delayed or restricted by a dominant old male are another.

Such cases, I need hardly say, never bother armor-plated neo-Darwinians. But then *no* cases, possible or even actual, ever do bother them. If you discovered tomorrow a new and most un-Darwinian-looking species of animals, in which every adult pair produced on average a hundred offspring, but the father always killed all of them very young, except one which was chosen by some random process, it would take an armor-plated neo-Darwinian no more than two minutes to "prove" that this reproductive strategy, despite its superficial inadvisability, is actually the optimum one for that species. And what is more impressive still, he will be able to do the same thing again later, if it turns out that the species had been misdescribed at first, and that in fact the father always lets *three* of his hundred offspring live. In neo-Darwinism's house there are many mansions: so many, indeed, that if a certain awkward fact will not fit into one mansion, there is sure to be another one into which it will fit to admiration.

IV

It is by no means true, then, even of all animal populations, that they are always as large, or are rapidly tending to become as large, as the available food would permit. For populations of pines, cods, and countless other species, it is no doubt a useful approximation to the truth, to say that they always blindly and quickly multiply up to the numbers that there is food to support. But by the time one gets to man, it is a grotesque travesty of the truth to say this. Human life is full of opportunities for reproduction which the supply of food would permit, but which are not taken in fact.

Consider the most familiar and omnipresent kind of human population: a family, consisting of a father, mother, and at least one son and daughter. If the Malthus-Darwin principle were true, as many offspring as there is food to support would always be produced not only by the father with the mother, but by the mother with each of her sons, and by the father with each of his daughters. Since this does not happen always and everywhere, the Malthus-Darwin principle is false. Whether in fact it has ever happened even once, in the entire history of our species, may very reasonably be doubted: so strong and general is the aversion to incest.

As well as being averse to incest, our species practices, or has practiced, on an enormous scale, infanticide, artificial abortion, and the prevention of conception. No other species does anything at all of this kind, but we do, and we appear to have done so always. If the Malthus-Darwin principle were true, then every human life which has ever been deliberately ended before birth or shortly after it, or has ever been deliberately prevented from beginning, would otherwise soon have been ended anyway, by starvation. Have you ever heard of anything more ridiculous than this? In the city where you live, hundreds or thousands of artificial abortions are performed every day. Is it really true that there is not enough food to support even one of these potential children into adulthood? Well, that is what the Malthus-Darwin principle says, anyway: which must be admitted to be a consoling doctrine, even if not a true one.

But our aversion to incest, and our devotion to such things as contraception, are not the only factors in our lives which bring about neglect of opportunities for reproduction which the supply of food would allow. There are several others. One of these is the scope we give, especially to males, for widespread and even exclusive homosexuality, which at some times and places is not merely permitted, but has a high value placed upon it. There could not possibly be any species which does this, if the Malthus-Darwin principle were true.

Another peculiarity of our species, which leads us to neglect opportunities for reproduction that the supply of food would have permitted, is this: that marital fidelity is generally enjoined, especially upon women. Of course this injunction is often violated. But then, it is also often obeyed. Is this something which you would expect to find among humans, any more than among pines or cod, if the Malthus-Darwin principle were true?

A further peculiarity of our species which has the same anti-reproductive effect is the fact that a high value is widely placed on virginity at marriage, while women are hardly ever permitted to marry as soon as they are capable of reproduction. The result is, of course, that *years* of reproductive opportunities are very commonly neglected, however plentiful food may be.

These, then, are some of the things which prevent, in our species, reproduction which the supply of food would have permitted: our aversion to incest, our positive measures for repressing the increase of our numbers, the scope we allow to male homosexuality, and the value we place on the fidelity of wives and on virginity at marriage. No doubt there are other such things. The result of them altogether is that even if the Malthus-Darwin principle is a useful approximation to the truth for pines, and cod, it is nothing of the sort for human beings.

Malthus, as we saw, advanced his principle of population for all species of organisms indifferently. His book could hardly have engaged the attention of naturalists, such as Darwin and Wallace, if he had not done so. But Malthus was not himself a naturalist, and had no interest whatever in general biology. He was interested only in man, and the purposes of his *Essay on Population* were altogether particular-historical, and even polemical. It is therefore a curious irony that the general biological principle which he put forward comes steadily closer to being true, *the further one departs from the human case*, and is a grotesque falsity only in the one case which really interested Malthus: man.

Human populations, once they reach a certain size and complexity, always develop specialized orders, of priests, doctors, soldiers. To the members of these orders, sexual abstinence, either permanent or periodic, or in "business hours" (so to speak), is typically prescribed. Here, then, is another fact about our species which is contrary to what one would expect on the principle that population always increases when, and as fast as, the amount of available food permits.

That priests, doctors, and soldiers sometimes violate their professions' prescription of sexual abstinence should go without saying. We must submit with whatever patience we can find to all the good old stories about the sexual behavior of nuns, monks, etc., which are told by Freudian and other sex maniacs, by neo-Darwinian reproduction maniacs, and by Enlightened persons generally. Some of these stories are perfectly believable, of course, though others are not. But the one story which is perfectly *un*believable is the story that the Malthus-Darwin principle tells; that priests, doctors, and soldiers always and everywhere reproduce up to the limit set by the availability of food.

There are, then, several large and permanent professions of people, from whom is required a greater or lesser degree of sexual abstinence. But as well as that, there is in human life the immense phenomenon of general (not necessarily professional) religious sexual asceticism.

There could not possibly *be* such a thing as religious sexual asceticism, of course, in our species any more than in any other, if the Malthus-Darwin principle were true. But that, again, is just too bad for that principle. The simple fact of the matter is that large and enduring religious communities, committed to complete sexual abstinence, and largely (at least) practicing it, are a constantly recurring feature of human history. In Western civilization, for more than a thousand years, there was no more important institution—important morally, economically, and culturally—than Christian monasticism.

Just how successful Christian monks and nuns were, overall, in living up to their vows of sexual abstinence, we cannot of course know. (We can and do know that the members of the great nineteenth-century American movement of the Shakers, for example, though the sexes were mixed in their communities, were almost entirely successful in living up to *their* vows of complete sexual abstinence.) But in any case, the precise level of chastity in Christian monasteries does not concern us here. What does concern us is that, if the Malthus-Darwin principle were true, then nuns and monks would have had, on the average, not only as many children as the secular clergy and the laity did, but as many as the food available was capable of supporting. No student of history, unless he happens to be also a raving lunatic, will believe this.

To neo-Darwinians, of course, as to all Enlightened people, the existence of religion is a fact completely and uniquely inexplicable. Even though their explanatory pretensions are boundless, neo-Darwinians have never yet essayed the "sociobiology" of *religion*, and are not likely to do so in a hurry. The temptation for neo-Darwinians is, rather, to "write off" religious sexual asceticism, as being admittedly inexplicable, but having at any rate the partially redeeming feature that it is an isolated case. But it is not an isolated case.

Religion is not at all the only thing in human life which has a marked tendency to repress or extinguish reproduction, and even to mortify the sexual impulse itself. Intense and prolonged thought, in the few people who are capable of it, has the same tendency. So does high artistic creativity. In fact either of these things is, in general, far *more* strongly and uniformly unfavorable to reproduction than religion in general is.

That men of intellectual or artistic genius are comparatively infertile is an old belief. It is also a well-founded one. But it has always both puzzled and annoyed those who adhere to the Malthus-Darwin principle. In fact one Darwinian—Francis Galton, no less—wrote a famous book, *Hereditary Genius* (1869), for the express purpose of refuting this ancient opinion. The book deserves all its fame, and more, but it completely fails, or rather it scarcely even attempts, to do what it was intended to do. Well, Galton's project was doomed from the start: this is a matter on which there can be no two opinions.

Among men of the highest genius, the twenty-two children of J. S. Bach are, of course, enormously exceptional. In fact, even Charles Darwin, with seven children surviving into adulthood, is a prodigy of fertility in this class. Indeed, even the number of Montaigne's children—"three or four," according to his own memorable report—is still well above the median value for this class. So is Shakespeare, with three children. Mozart is closer to it, with two, but is probably still above it. For what depresses the average fertility of this class beyond all hope of recovery is the huge over-representation of the childless. Men of the highest genius who were childless include Newton, Faraday, and Mendel; Vivaldi, Handel, and Beethoven; Gibbon, Macaulay, and Carlyle; Plato, Aquinas, Bacon, Locke, Leibniz, Hume, Kant and Mill. Anyone who thinks he can frame a list of comparable individuals who had on the average enough children to counter-balance the childlessness of these, is welcome to try; but he will not succeed. No rational person will suppose that this association of extremely low fertility with the highest intellectual or musical genius is accidental. Still less will any rational person suppose that the failure of the childless great to achieve parenthood was due to shortage of food. Whatever it was that prevented Newton or Handel or Kant from ever even copulating with a woman, it was certainly not hunger.

What is true at the very top of the scale of intellectual or artistic gifts continues to be true, all the way down that scale. Just below the very highest level, an average number of children is again far below the average for people of no intellectual or artistic distinction. And it is, again, spectacularly depressed by the huge contribution of the childless: Copernicus, Swift, Adam Smith, Samuel Johnson, Haydn, Dalton, Francis Galton himself—to mention a few examples at random. And so it goes on, with the average number of children going slowly up, as the degree of intellectual or artistic ability goes slowly down. But the lives of even average philosophers, scholars, scientists, or composers are, and always have been, a sufficient refutation of the Malthus-Darwin principle. Any fair comparison, between the average number of children that *they* have and the average number that other kinds of people have, will reveal that they are even further from satisfying the Malthus-Darwin ideal than those other people are.

BUT NOW let us set aside the influence on reproduction of intellectual or artistic gifts, of religion, and even of membership of the priestly or medical or military profession. Think of people whose actual reproductive careers are altogether unaffected by any of those things. Even these people, I venture to think, have probably *never in a single instance* had as many children as there was food to support. Of course I cannot prove this. Still, it is reasonable to believe it. For even without the impediments due to high culture, religion, or a profession, the restraints which are placed on reproduction by the aversion to incest, by the ease with which infants can be killed or aborted or prevented, by the scope allowed men for homosexuality, and by the prescriptions imposed on women in favor of virginity before marriage and fidelity after it, are together so very great that it is reasonable to doubt whether any human being has ever *completely* escaped their practical influence.

If this is so, then every human being who has ever lived is, in fact, a sufficient refutation of the Malthus-Darwin principle. But even if I am only nearly right—if only one, or ten, or only a million people have ever had as many children as there was food to support—then *nearly* every human being who has ever lived is a sufficient refutation of that principle.

V

Darwin must have gone wrong somewhere about man, and badly wrong. For if his theory or explanation of evolution were true, there would be in every species a constantly recurring struggle for life: a competition to survive and reproduce which is so severe that few of the competitors in any generation can win. But this prediction of the theory is not borne out by experience in the case of man. In no human society, whether savage or civilized, is there any such struggle for life. At least, no such struggle is anywhere observable. This is the inconsistency, between Darwin's theory and the observable facts of human life, which I called in the preceding essay "Darwinism's dilemma."

In the present essay (unless I in turn have gone badly wrong somewhere), I have pointed out where it was that Darwin *first* went wrong about man. It was in accepting Malthus's principle, that in every species population always presses on the food available, and tends to increase beyond the size that there is food to support. This principle is not true without exception even in the case of other species, but in the case of our own it is extravagantly wide of the truth. Until he embraced this principle in 1838, Darwin was not logically committed to any propositions about man which are false. But once he adopted, in an attempt to *explain* evolution, the principle which Malthus had put forward, he *did* logically commit himself to a false belief about man.

Darwin's explanation of evolution, then, contains as an essential element a proposition which is false in the case of man. This conclusion is plainly of some historical interest, if it is true. But it is also of theoretical interest. For it means that Darwin's explanation of evolution, even though it is (as I said earlier) still the best one available, is not true.

Essay 3
"But What About War,
Pestilence, and All That?"

The primary or fundamental check to the continued increase of man is the difficulty of gaining subsistence and of living in comfort. We may infer that this is the case from what we see, for instance, in the United States, where subsistence is easy, and there is plenty of room. If such means were doubled in Great Britain, our number would be quickly doubled.
—Charles Darwin, *The Descent of Man*

T O MANY PEOPLE, the preceding essay will seem to be exposed to an easy and fatal retort, which could be expressed as follows:

Darwin and Malthus did *not* think that organic populations are always as numerous as their food supply permits. That would amount to believing that the limitedness of food is the *only* check to population. But Malthus, as is well known, said that human increase is restrained not only by "famine," but by war, and also by "pestilence" (i.e., disease). He further pointed out yet another check to human population, which he called "vice": that is, such things as contraception, infanticide, artificial abortion, and homosexuality. As for Darwin, he knew as well as anyone has ever known that the increase of organisms is repressed, not only by shortage of food, but by predation and disease.

47

This objection is true. Well, of course it is! How could people like Darwin or Malthus *not* have known that human numbers are checked by war and infanticide (for example), or that animal and plant numbers are checked by disease and predation? In their time as in ours, it would be a very ignorant person who did not know these facts. And it is perfectly true that Malthus and Darwin not only knew them, but often stated them in print.

Does the preceding essay, then, rest on a misrepresentation of Darwin and Malthus? Were my criticisms wasted on mere men of straw? No. The reason is this: that although Darwin and Malthus often acknowledged the existence of checks to population other than limited food, they also believed that all these other checks are of negligible importance, compared with the check imposed by limited food. In other words they believed that, as near as makes no difference, the size of an organic population *does* depend only on its supply of food.

The proof of this is that, almost invariably, Malthus and Darwin say in effect, "population increases if food does," and even "population increases as a simple monotonic function of any increase in food": flatly, like that. They almost never say, "Population increases if food does, unless it is prevented from doing so by an outbreak of disease," or "population doubles if food doubles, unless it is prevented from doing so by an increase in predation." Almost always, they simply leave out all such qualifications or, if they do insert some of them, they do so only in order to dismiss these other checks to population as of negligible importance.

It is certainly very tiresome for an author to have to say "other things being equal," or "if we suppose other factors unchanged," every time that such a proviso is needed to make what he says strictly true. This proviso, if constantly inserted, also tires the reader, and thus adds to the obstacles—always plentiful enough —to the writer's conveying to the reader's mind just what he intends to, and nothing else. So there is often some excuse for an author's omitting a certain necessary qualification to what he says. And his omission of it will be not only excusable but fully justified, of course, if he has made it clear in the immediate context that he does intend this qualification to be understood.

But these general literary considerations go very little way towards explaining the neglect, by Malthus and Darwin, of checks to population other than the limitedness of food. That neglect is far too systematic, and far too absolute, to be explained by a desire to keep the number of "other things being equal" clauses within bearable limits. It is so systematic an absolute, in fact, that it can only be explained by supposing that Malthus and Darwin *believed* all other checks to population to be of negligible importance, compared with the check imposed by limited food.

Some examples of this neglect follow.

Malthus says, of "nations of hunters" such as the North American Indians, that "their population is thin [i.e., sparse] from the scarcity of food," and that "*it would immediately increase if food was in greater plenty.*"[1]

Nor is there anything in the context of these words, any more than in the words themselves, to the effect of ". . . unless it were prevented from doing so by a desolating epidemic or war."

It is not only when Malthus is dealing with special human groups, such as nations of hunters, that he absolutely neglects checks to population other than the scarcity of food. He does so equally when he is stating a general proposition about human population. He says, for example that "population does invariably increase when the means of subsistence increase,"[2] and that "population constantly bears a regular proportion to the food that the earth is made to produce."[3] That is to say, not only "population increases if food does," but "population increases as some simple monotonic function of any increase in food." And once again, there is nothing in the context where the above statements occur to the effect of ". . . unless it is prevented from doing so by disease, or war, or vice, or some combination of those things."

Here is Darwin writing in exactly the same vein as Malthus:

The primary or fundamental check to the continued increase of man is the difficulty of gaining subsistence and of living in comfort. We may infer that this is the case from what we see, for instance, in the United States, where subsistence is easy, and there is plenty of room. *If such means were doubled in Great Britain, our number would be quickly doubled.*[4]

The reference here to "room"—i.e., space—can safely be ignored. Limitedness of space is quite certainly not a check to population at all, in anything like the same sense as limitedness of food is, or disease, or predation, or war. To suppose that it is would be too reminiscent of the old Jewish preacher, lamenting the great and apparently endless sufferings of his nation, who concluded thus: "For a Jew, it would be better not to have been born; but scarcely one in a hundred is so lucky." An organism is certainly unlucky, if it can get no food to put in its stomach; but an organism which can get no space to put itself in, is not *unlucky*—just nonexistent. In any case, Darwin would have known perfectly well that the area of Great Britain could easily have supported twice its population of 1874 (the year he published the statement above), if enough food for them had been available.

What Darwin says in the above passage amounts, therefore, to this: if food were doubled in Great Britain, population would quickly be doubled. Then, the reference to Great Britain was obviously merely illustrative: any other country would have done equally well. So that what Darwin was really asserting was the quite general proposition, that population doubles if food doubles: as flat and unqualified as that.

It is true that Darwin in this case was not quite so incautious as Malthus had been in the passages I have just cited. For he does mention, later in the same paragraph, some checks to population other than limited food. But he mentions them only in order to dismiss them as negligible. Thus he writes that, "the effects of severe epidemics and wars are soon counterbalanced, and more than counterbalanced, in nations placed under favourable conditions. Emigration also comes in aid as a temporary check, but, with the extremely poor classes, not to any great extent."[5]

A reader could very reasonably ask, why does Darwin not consider the effects of epidemics and wars on nations *not* placed in favorable conditions? And why does he not mention the effect on population of emigration by those who are *not* extremely poor? Again, why has he here made no mention whatever, even in the brief and euphemistic manner of his time, of those potent checks to human increase which Malthus had called "vice"?

No, there are, in Darwin and Malthus throughout, just too many of these things: *omissions of any reference* to checks to population other than limited food, *mentions of only some of those checks*, and *dismissals as unimportant* of those other checks when they *have* been granted a mention. It is impossible to explain all these facts, except by recognizing that Darwin and Malthus believed that the size of organic populations *does* depend only, or near enough only, on "the difficulty of gaining subsistence." Wallace spoke for them, as well as for himself, when he wrote that "a constant supply of wholesome food is *almost the sole requisite* for ensuring the rapid increase of a given species. . ."[6]

What I cannot possibly prove here, of course, is that the four passages I have just quoted are *typical* ones. That is, that Malthus and Darwin almost invariably write, as they do in these passages, as though the limitedness of food is the only check to organic increase which is worth mentioning. But anyone who hopes to show that my quotations are untypical ones will need to begin by matching them with four similar passages, in which a restraining effect on human numbers, comparable with the effect of limited food, *is* ascribed by Darwin or Malthus to war, or to disease, or to "vice," or to some combination of those checks. I will venture to affirm that there is not one such passage anywhere in Darwin or Malthus. I am much more confident still, that there are not four of them.

The fact to which I have just indirectly drawn attention, that all of my four quotations were about *human* population, is mere happenstance. It does not affect, either way, the value of these quotations as examples of the neglect by Malthus and Darwin of

DARWINIAN FAIRYTALES

checks to population other than limited food. But it may help to remind us that, where it *is* our species which is in question, the proposition that population increases if food does is in need of qualifications so numerous as to make it entirely uninteresting.

There are, first, the qualifications which Malthus and Darwin themselves made to it (and then effectively ignored). That is, "population increases if food does, unless it is prevented from doing so by disease, or by war, or by emigration, or by homosexuality, or by contraception, abortion, or infanticide." But then, there are many other qualifications which are equally necessary, even though most of them were completely undreamt of by either Malthus or Darwin. For example, ". . . unless there are widespread massacres, persecutions, or deportations." Then, ". . . unless there is a mass revival of sexual asceticism." Again, ". . . unless a suicide cult sweeps through the population." Yet again, ". . . unless there is an epidemic of feminist motherhood-phobia."

This list of necessary qualification is already very long. Yet no person of common sense will suppose that it is complete, or will suppose that anyone knows how to complete it. The Malthus-Darwin proposition, then, that population increases if food does, may be a truth, or a false but fertile near-truth, when it is applied to species other than *Homo sapiens*. But applied to our species, the best it can be is the following pure triviality: that population increases if food does, unless it is prevented from doing so by one or more of a dozen different causes that we know of, or by one or more of an indefinite number of causes that we do not know of. "For this relief, no thanks."

But as a matter of historical fact—as I said in the preceding essay, and have tried to substantiate in this one—Malthus and Darwin in effect maintained, with *no* qualification of any importance, even as applied to man, that population increases if food does. In other words, that it is scarcity of food alone which limits organic increase. And then, what *can* one say of this Malthus-Darwin "principle of population," except that, as applied to man at least, it is false, or rather, ridiculous? Which is what I said in the preceding essay.

EVEN OUTSIDE the human case, that principle is false. Animal populations are often not as large as their food supply would permit (as we saw in the preceding essay). Even for plants the proposition that population increases if food does stands in obvious need of qualification, to the effect of ". . . unless there is an increase in disease or predation or some other check."

And yet there *does* seem to be something about the Malthus-Darwin principle which is *broadly* true, and profoundly important. Among the various checks to population, scarcity of food seems somehow to stand apart from all others, and to be of unique importance. On the people who observe them, animal and plant populations make the general impression, at least, that whenever scarcity of food ceases for a while to restrain their tendency to increase, all other checks to population are found barely sufficient, even when combined, to prevent a large actual increase in their numbers.

This impression is, I believe, well-founded. But it is hard to turn it, from an impression, into a proposition which is not obviously false and yet is definite enough to be worth saying. It certainly will *not* do to say, for example, that the check to population from scarcity of food is at all times *stronger than* all other checks put together. For that would imply that the extinction of the dodo was due to food shortage more than to anything else; whereas it was in fact due entirely to human predation. There have also been other cases, presumably, in which a species has been extinguished amid plenty of food.

Yet there is a certain respect in which food scarcity does stand quite apart from such checks to population as disease and predation. This respect will be more visible if we transpose the Malthus-Darwin picture from negative to positive (as it were) and instead of thinking about checks to population, ask ourselves, what is the struggle for life within a species a struggle *for*? Population is kept within bounds by the losers in that struggle meeting early death from starvation, predation, disease, or whatever. But what is it that the winners get, and which conspecifics compete with one another *to* get?

Well, no doubt, they compete with one another for some benefit, or some advantage over the other competitors. And yet, not for just *any* kind of benefit or advantage. For some of the advantages or benefits which can accrue to an organism are completely "sharable": that is, A's having or getting this advantage does nothing whatever to prevent B's having or getting it. And it makes no sense to speak of A and B struggling or competing with one anther for X, if X is something that A's getting is no obstacle at all to B's getting too.

One example of a completely sharable advantage is immunity to a certain disease. If organism A has this immunity, while some of its local conspecifics do not, this fact will certainly give A an advantage (if other things are equal) in the struggle for life. Yet A's having this immunity does nothing to prevent B's also having it. It is not as though there is only a limited fund of the immunity to be shared out among the members of a species, so that more if it for one means less for another.

Another completely sharable advantage is improved defense against predators. Suppose that A has better defense against predation than any of its ancestors had, and better than some at least of its contemporary conspecifics have. (The improved defense might be better camouflage, sharper hearing, or whatever.) This fact will certainly give A an advantage (if other things are equal) over some of its rivals in the struggle for life. Yet A's having sharper hearing (say) does nothing at all to prevent any conspecific B from also having sharper hearing. Here again, more for A does not mean less for B.

But there are other benefits which are completely *un*sharable: A's getting the benefit precludes B from getting it. Food is the most obvious example. If I get the benefit of a certain bit of food, you do not. In many species there is a certain exception to this, of course, when females are pregnant, or are engaged in early feeding of their young. But with that exception, food is a benefit which cannot be shared.

The struggle for life which, according to Malthus and Darwin goes on in every species, *cannot*, therefore, be a struggle for such things as immunity to a certain disease, or improved defense against predators. It can only be a struggle for benefits or advantages which are completely unsharable, as food is. So—if we now switch back again from positive to negative and think, not about the benefits winners get, but about the checks which weed out losers—we see that Malthus and Darwin had, or anyway could have had, a very good reason for putting the difficulty of getting food in quite a different class from such checks as disease and predation.

That is not to say that they were justified in virtually neglecting, as they did, all checks to population other than the scarcity of food. Nothing could justify the length to which they carried that neglect. And yet even that neglect is more understandable than I have so far indicated.

For what if *all* the benefits or advantages which can accrue to an organism, except food, were of the completely sharable kind? That is, were things like immunity to a disease, or improved defense against predators. In that case food would be the only benefit which is completely unsharable. It would therefore be the only possible object of a struggle or competition for life among conspecifics. And then, losing in the competition for food would be the only way of losing in the struggle for life; in other words, scarcity of food would be the only check to population.

Now (continuing to set aside the human case), surely that supposition corresponds extremely closely, at least, to the actual case? For what completely unsharable benefits or advantages are there that an organism can enjoy, except food?

Air? It is certainly unsharable enough (supposing we agree to call it a benefit at all). The air that fills my lungs cannot also fill yours (with the exception, again, of pregnant females). But what disqualifies air from being a benefit coordinate with food is that, during the time in which terrestrial evolution has taken place, it has never been scarce enough to become an object of competition or struggle among organisms.

A priori, of course, it could have been, and perhaps it once actually was. If life on earth were confined to the plants and animals contained in one tiny pocket or jar of air, then superior equipment for air intake would presumably be just as advantageous to an organism as superior equipment for food intake would be. But that supposition is just too remote from the actualities of terrestrial evolution to be at all enlightening.

Even in that one little jar of plants and animals, however, the parallel between air and food is very imperfect. It would be true to say, of an animal species represented in this jar, that its number will increase if its food does, if all other checks to increase remain the same. But would it be true to say of this species that its number will increase if its *air* does, all other checks remaining the same? Clearly not, and the reason is obvious. There is a plain causal chain leading from an animal's food intake to its nutrition, thence to its growth, hence to its reproduction, and thus to its population increase if other checks do not prevent that final result. Indeed, *every* link in this chain can be prevented, by any one of a thousand contingencies, from taking place. But there is no corresponding causal chain at all which leads from an animal's air intake to its population increase.

Apart from air, I have not been able to think of any even momentarily plausible candidates for the role of a completely unsharable benefit other than food. Hence I am unable to suggest what a struggle or competition for life among conspecifics could possibly be a struggle or competition *for*, except food.

My last few pages therefore constitute (unless they contain some mistake) a kind of justification of the neglect by Malthus and Darwin of checks to population other than the limitedness of food. More generally, they constitute a justification of all the countless nineteenth-century Darwinians who could not be bothered distinguishing clearly and consistently between the "struggle for *life*," and "the struggle for *the means of subsistence*." Darwin himself was the earliest of these Darwinians, and all the later ones simply followed his example.

But my justification is only of a conditional kind, and is two-edged. For what it comes to is this: "scarcity of food *is* the only check to the increase of a species, *insofar as* conspecifics are engaged in a struggle or competition with one another to survive." And while this can obviously be read as a justification of Malthus and Darwin, it can equally well, or rather, better, be read as a *reductio ad absurdum* of them. For we know, after all, that scarcity of food is *not* the only check to organic increase.

Malthus and Darwin knew it too, of course. As I said at the beginning of this essay, they were as well aware as we are of predation and disease as checks to population. They mentioned them often enough, too. But Malthus and Darwin were held captive by the picture of a struggle or competition or battle for life among conspecifics, while disease and predation simply will not fit into that picture. As we have seen, you *cannot* struggle with a conspecific for immunity to a certain disease, or compete with him for improved defense against predators. And so, perforce, disease and predation simply had to fade out of the Malthus-Darwin picture, leaving food as the only possible thing for all the struggling to be about.

If the reader has found the latter half of this essay a little tortuous, I can at least plead in defense that it is as plain as a Roman road compared with Chapter III of *The Origin of Species*, entitled "Struggle for Existence." No chapter of the book was more critical for Darwin's whole argument than this, and yet no other chapter is more, or even equally, bewildering. There are several references in it to predation and disease as checks to population, and yet no reader could possibly carry away from this chapter a single clear idea as to what, except food, the struggle for existence could be a struggle *for*.

Population, Privilege, and Malthus' Retreat

[If Malthus (and Darwin) had been right in thinking that population is restrained principally by the difficulty of getting food], then the English would have long ago become a *people of nobles*.
—William Godwin, *Of Population* (1820)

MALTHUS' *Essay on Population* of 1798 was an anti-communist and anti-socialist tract. It claimed to point out a fatal flaw in all proposals for abolishing private property, or for equalizing wealth. The *words* "communism" and "socialism" do not occur in it, but that was simply because those words did not exist when Malthus wrote. Where we would speak of communist or of socialist political programs, Malthus spoke of "systems of equality": an expression which goes rather more to the heart of the matter (when you come to think of it) than do the expressions we use.

Schemes for community of property or for the equalization of wealth had been pouring out of France for fifty years when Malthus first published his *Essay*. They came from the pens of Mably, Rousseau, and Morelly, among others. In the 1790s such schemes had been powerfully advocated in France by Condorcet and Baboeuf, and in England by William Godwin in *Political Justice* (1793), and Thomas Paine in *The Rights of Man* (1792). But

Malthus was convinced that communism would replace the existing *comparative poverty of most* by the *absolute poverty of all*, and that it would, in the process, destroy "everything which distinguishes the civilised from the savage state."[1]

He was convinced of this, both by an argument from his principle of population, and by certain economic arguments which do not depend on that principle at all. These economic arguments also convinced him that the system of "Poor Laws" which existed in England in his own time—that is, the system of publicly-funded unemployment relief—was already a long step towards socialism, and hence towards economic and cultural disaster.

For every one person who has actually read Malthus's *Essay*, there are a hundred people willing to talk or even write about it. This has always been the case, and has often been remarked upon. I do not know why it is so, but one effect which could easily have been anticipated from it has certainly taken place. This is that ridiculous misconceptions about the book become widely and firmly entrenched in people's minds, to a point where it is quite hopeless to dream of ever dispelling them.

One of these misconceptions is that Malthus's *Essay* advocates, either openly or covertly, the practice of contraception. Nothing could be further from the truth than this belief. Malthus was fiercely opposed to contraception, and made this fact sufficiently clear in his book. Yet, on no other foundation than this ludicrous error, a new word came into existence early in the nineteenth century, and remained current for a hundred years: "neo-Malthusian," which meant (when applied to things or practices) "an aid to contraception," and (when applied to persons) "an advocate of contraception." (I should perhaps add that the word "contraception" itself has existed only since 1917). And though "neo-Malthusian" is no longer in use, almost everyone at the present day who is educated enough to have heard of Malthus at all "knows" that he was the great apostle of contraception, the St. Paul who brought this saving grace into modern life.

But an even more grotesque misconception about the *Essay* is one of the achievements of the twentieth century. This is the belief that Malthus, with wonderful prescience, had written his book in order to warn humanity of the catastrophic "over-population" which was even then impending over us, and which is now—because we have failed to heed his warning—about to descend upon us. There are literally millions of people nowadays who believe this, even if they believe nothing else about Malthus.

Yet even someone who had never read Malthus ought to be able to work out that this belief cannot possibly be true. All that such a person would need to know is that Malthus's book had supplied Darwin and Wallace with an essential component of their explanation of organic evolution in general. For suppose it were true, and that Malthus had said, that overpopulation threatens our species with universal famine: how could *that* fact have thrown any explanatory light whatever on the evolution of species from other species? An imminent halving or extinction of our species by starvation would undoubtedly be of practical interest to biologists, as to everyone else; but it is of absolutely no interest from the point of view of general biology. Darwin and Wallace could never have got a vital clue from Malthus, if he had been merely a fore-runner of the foolish or ignorant writers of the present day who try to spread panic about "over-population."

In fact this "catastrophist" interpretation of Malthus manages to be just about the exact opposite of the truth. His principle of population is a proposition concerning, not only all species of organisms indifferently, but all *times* indifferently. It not only does not predict any particular "crisis," or other journalistic artifact, in the history of humans and their food: it does not pick out any singularity in time at all in the history of any species. Quite the reverse, in fact: for it says that the relation between a species and its food is, in a certain respect, *always the same.*

The principle is this: that every population of any species is at all times as large as the available food allows, except when it is rapidly approaching that limit after having suffered a check from disease or some other cause, or after recently arriving in a new territory. At those times, the natural tendency to increase is less restrained than it normally is: new supplies of food, or a recent abnormally high proportion of deaths, provide an opening for an abnormally high proportion of a new generation to survive. Thus population "oscillates" (as Malthus is always saying) back to its normal size: that is, the maximum that there is food for.

This is evidently a theory which, so far from predicting any crisis or catastrophe concerning food and population, positively *excludes* such a thing, for any species. It is rather what you might call a "steady state" theory of population and food, or better still a "permanent plenum" theory. Of course it does not rule out the occurrence of famines, which may be of any given degree of severity and extent. But a famine is simply a period in which, from insufficient or inferior food, many members of a given population suffer death or debility at an earlier age than they otherwise would. That there can never for long *be* more people (or flies or whatever) than there is food for, should go without saying, and certainly needed no Malthus to teach us; though one sometimes suspects that our present-day population catastrophists (who imagine they are Malthusians!) believe, precisely, that there *can* be. What Malthus said was the far more interesting proposition (whether or not it is true) that there can never for long be *fewer* people, *either* (or flies or whatever), than there is food for. In other words, the tendency of organisms to increase is so strong that it neglects no opportunity to turn food-for-a-possible person (or fly or whatever) into food-for-an-actual one. Different species have, of course, very different gestation periods, and in most species sexual activity is confined to a certain season; but, subject to these and the like obvious qualifications, population increases *immediately* food does, and *exactly as much as* the increased food allows. *That* is what Malthus thought.

Yet most populations of organisms, most of the time, in spite of this *tendency* to exuberant increase, in fact increase only slowly or not at all. What, then, are the restraints or checks to population which operate effectually most of the time? In the human case, Malthus replies, the main ones are misery and vice. Human misery has, he says, three principal causes: famine, war, and pestilence, or—in less florid English—food shortage, war, and disease. By "vice" Malthus meant chiefly the use of "improper arts"[2] to prevent pregnancy or terminate it artificially; the "barbarous habit"[3] of infanticide; and "unnatural vice"[4] (i.e., homosexuality).

The misery check is common to all species whatever. No organism every willingly surrenders its life to hunger or disease or an enemy. War is peculiar to man (or nearly so) but large-scale killing, in the form of predation, is part of the fate of most species of animals and plants. The vice check, by contrast, Malthus says, is peculiar to man. Darwin agreed. The "instincts of the lower animals are never so perverted as to lead them regularly to destroy their own offspring . . .";[5] and neither, even if they *were* as perverted as to want to prevent conception, do they have the intelligence needed to succeed in doing so. According to Malthus, the vice check is even peculiar to *civilized* man: "vice [is] out of the question among savages."[6] Here Darwin demurred and said, what is true, that Malthus had underestimated the prevalence of infanticide and abortion among savages.[7] Neither man betrays any awareness of male homosexuality outside civilization, or of female homosexuality anywhere.

Such, in outline, was the theory of population and the checks to it which Malthus maintained in his *Essay*, and in which Darwin and Wallace detected a mechanism that would explain how species originate from other species.

THE MALTHUS-DARWIN principle of population (I have already said in earlier essays) is not true, at least with respect to man. But the principle has exercised so enormous an influence on biological thought that it will be worthwhile to expose it to a criticism which, though it again concerns the human case, is a little more subtle than anything that I have said against it so far. The criticism to be advanced in the next few pages is old, having been voiced at intervals during nearly two hundred years. But, for some reason which I do not understand, it has never commanded anything like the attention which, to me at least, it seems obviously to deserve. It concerns the association, or more accurately the lack of association, between fertility and privilege.

It clearly follows from Malthus's theory that, if there are two human populations in which vice is equally prevalent, one will be more fertile than the other if it is less miserable than the other: that is (according the Malthus), if it is less exposed to the misery resulting from food shortage, disease, and war. Or we may say the same thing in terms of one population, at two different times: supposing vice equally prevalent at both the earlier and the later stage, if the population suffers less misery from war, disease and food shortage at the later stage than it did at the earlier, it will then be more prolific of children than it was earlier. In plain English: other things equal, and on the average, people who are less miserable (or more privileged) have more children than people who are more miserable (or less privileged).

But this is not at all what we find in fact, either in history or in our own observations of everyday life. It is more nearly the very opposite of it. The words of a vulgar American song of the 1930s, that

The rich get rich
And the poor get children

come closer to expressing the uniform experience of mankind, than does this consequence of Malthus' theory.

It is quite certain that, in general, the poor are more exposed than the rich to the misery which results from food shortage, war, and disease. They are also less prudent and forward-looking than the rich. They are more ignorant, too, and therefore less able to make use of such prudence and foresight as they do possess. They have a smaller variety of things to occupy their minds than rich people do, and fewer sources of happiness open to them. They live more crowded together than the rich . . . Who could not easily continue this catalogue of differences? And then, we recall, sexual intercourse is one of the very few sources of happiness to which poverty is no bar. From all these familiar facts, common sense tells

63

us imperatively to expect, what we find to be actually the case, that large families are commoner, on the whole, among the poor than among the rich. In other words, that increase of population is more repressed, *not* where food shortage, war and disease fall more heavily, but precisely where they fall more lightly.

Of course, the distinction between rich and poor is not *exactly* the same as the distinction between the privileged and the rest: membership of a privileged class need not be constituted by *wealth*. It may be constituted by inherited rank, by individual military prowess, by religious authority, or by various other things. By any standards, and certainly by Malthus's standards, the Knights Templars in the year 1250, or the Jesuits in 1700, were highly privileged people. Few people at the time were less likely than they were to die of starvation or disease, or to be killed in war. But they were not *rich*, in the sense of having individual command of unusual wealth. Officially, indeed, and certainly in at least very many cases in fact, they had no money at all.

Of course the people who a little later *expropriated* both those privileged orders were convinced that their victims had left behind them, buried somewhere, mountains of gold. But that is merely a characteristic delusion of unscrupulous secularists who are short of cash. Even those expropriators, however, though like all enlightened persons they believed the worst about the official chastity as well as poverty of their victims, were never so stupid as to believe that the Templars or the Jesuits must have left behind them *offspring* proportional in numbers to the degree of privilege which the members of those orders had enjoyed. But that is what they should have believed, and what should have been true too, if Malthus's theory were true.

Go to the extreme case and consider the *most* privileged classes of people that history can show: the people for whom the probability of death from starvation or in battle was lowest, and to whom the best medical attention of the time was available. Such classes have *never* been prolific of offspring in anything like the degree to which they were privileged. They have never even managed to *maintain* their numbers by reproduction. They have survived, when they have, by early recourse to non-biological expedients: recruitment or adoption.

The offspring of a most privileged class exhibit, in fact, more strongly than those of any other class, and far more strongly than the offspring of the poor, a proclivity towards a whole range of things, every one of which is more or less unfavorable to parenthood. To early sexual exhaustion, to sexual incapacity, to sexual indifference, to homosexuality, to religion, to study, to art, to connoisseurship, to gambling, to drunkenness, to drugs . . . To almost anything in the world, in fact, except increasing or even maintaining the numbers of their own class by reproduction.

There is in Malthus's favor, of course, a certain tendency which privileged men have, to leave more children than unprivileged ones do; for the obvious reason, that they find more women sexually accessible. But there are also other and opposite tendencies which, singly or in combination, for the most part prevent that tendency from achieving very pronounced expression. One such tendency is the higher probability of promiscuous men contracting a sexually transmitted disease inimical to parenthood. Another is the tendency that privileged women have to leave fewer children than unprivileged ones; partly, no doubt, for the equally obvious reason that they are more difficult of sexual access. Any theory which predicts that exceptionally privileged women, such as Cleopatra, Elizabeth I of England, Catherine the Great of Russia, Queen Victoria, and the actress Elizabeth Taylor, will on the average have exceptionally many children would be contrary not only to fact but to common sense. But that is what Malthus's theory predicts. (Those five women in fact averaged, so far as I can learn, about four children each: certainly not an exceptionally large number.)

Again, we can easily concede to Malthus that a large part of the infertility of the privileged is due, sometimes at least, to their stronger propensity to "vice" in the form of contraception or abortion. But here too there are countervailing tendencies. We are apt to be misled, on this subject, by thinking only of middle-class people, in countries like our own during the last hundred years. They, indeed, have lived under a heavy and increasing cloud of anxiety about the number of children that they can (as they vulgarly say) "afford." But a rich person is more or less exempt from that anxiety; while an aristocrat despises it, just because it *is* a middle-class anxiety. And not only the economic but even the physical burdens of motherhood are far lighter for *highly* privileged women than for others. Once the baby is actually delivered, *the staff* take over the whole of that not-very-interesting business.

The Malthus-Darwin principle of population, then, when it is applied to man, not only fails to predict the right relation between fertility and privilege, but predicts what is roughly the inverse of it. The incidence of "vice" in Malthus's sense varies in the wildest manner from time to time and from place to place. No sane person will believe, for example, that abortion and contraception were about equally prevalent among the Catholic ruling class of Austria in 1570, the Puritan theocrats of New England in 1680, the Whig lords who ruled England in the mid-eighteenth century, and the Japanese imperial circles of 1935—to take a handful of cases at random. But there is one fact which does emerge from human history with unvarying insistence, and it is a fact which is fatal to the Malthus-Darwin theory: that the natural rate of human increase is repressed the more, *not* where the misery due to famine, war, and pestilence falls more heavily, but precisely where it falls more lightly.

It is always painful, as Huxley said, to witness the brutal murder of a beautiful theory by an ugly fact. The Malthus-Darwin theory of population is certainly a beautiful theory, partly because it comprehends, in one simple biological scheme, the human and all other cases. Nevertheless, and whether or not anyone wants to know, it was brutally murdered, ages before it was born, by an ugly fact about human beings.

VARIOUS AUTHORS have made the kind of criticism I have just been expounding of the Malthus-Darwin theory of population. The most recent one I know of (though it would be surprising if he were really the most recent) is R. A. Fisher, in *The Genetical Theory of Natural Selection* (1930).

Fisher had long been puzzled by the question—a very reasonable question too, if you start off by looking at man from a Darwinian point of view—why it is that successful civilizations do not just go on being even *more* successful. Why is it that, instead, they all sooner or later succumb to some less advanced rival? Darwinian biology affords no non-human parallel to this. And Fisher's answer, given in the last five chapters of his book, was as follows. First, that comparative fertility or infertility is to a considerable extent inherited; and second, that advanced civilizations have always practiced a systematic "social promotion of the less fertile."[8] As a result, they have always experienced a general decline in ability, since the very people whose abilities qualified them to rise in society have always been subjected to a selective pressure to have few or no children.

This theory, whether it is true or not, is at least extremely plausible, and as depressing as it is plausible. Yet it did not lead its author entirely to despair. For Fisher believed that it might yet be possible to create an advanced civilization which is permanent and progressive, by the adoption of certain eugenic measures. In particular, financial encouragements, increasing with the number of their children, should be given to people who—well, to be brief, to people like the author of *The Genetical Theory of Natural Selection*.[9]

The germ of Fisher's theory, as he acknowledged,[10] had been supplied by Francis Galton, in a brilliant piece of historical and statistical detective work nearly seventy years earlier. The question had been, why do British peerages—that is, their direct male lines—expire with the extraordinary rapidity that they do? Galton's answer, published first in an article in 1865[11] and later in his book *Hereditary Genius* (1869), has never (as far as I know) been successfully challenged. It was, that peers have a fatal tendency to marry *an heiress*: someone, that is, whose parents had not managed to produce even one son, and who is herself, therefore, likely to inherit her parents' comparative infertility. It was this markedly non-eugenic propensity of peers, or rather this violently *dysgenic* propensity, which led Galton to call the peerage a "disastrous institution."[12]

Galton was by no means the only person in the 1860s who saw the Malthus-Darwin theory being refuted before his very eyes: who saw, that is, that the people least subjected to food shortage, war, and disease, and least prone to economic anxiety about the number of their children, were reproducing at the lowest rate, instead of at the highest, as Malthus's theory predicts. One of the others was a writer now entirely though un-deservedly forgotten, W. R. Greg. He published criticism of Mal-thus and Darwin along the lines I have indicated, first in an article of 1868[13] and later in his book *Enigmas of Life* (1883). Another was no less than A. R. Wallace, who had published ar-ticles along similar lines as part of his disagreement with Darwin about man and natural selection. But I have not myself been able to meet with those articles.[14]

THE HONOR of having originated this line of criticism belongs, however, to William Godwin. To the very man, that is, against whom more than any other Malthus had originally written his *Essay*, and over whom he is generally considered to have enjoyed a complete triumph. In 1820, Godwin published *Of Population*, which was his long delayed major reply to Malthus. The book was a complete failure, and certainly did not, as a whole, deserve to succeed. But it deserved still less to fail as a whole, because it contained a number of "palpable hits." One was Godwin's remark that if Malthus's principle of population were true, then the English would long ago have become—as they certainly have not become—*"a people of nobles."*[15] Here, for once, Godwin was concise when he should rather have been copious, instead of the other way about. He may fairly claim to have put Galton, Greg, and Wallace, and even more R. A. Fisher, to shame for their long-windedness.

As I said earlier, the line of criticism of which I have been speaking has never received the attention it deserves. It ought to have compelled an early and public admission, by Darwin and all other interested persons, that our species, at any rate, does *not* conform to Malthus's principle of population. But nothing in the least like that has ever happened. Have you ever so much as heard before of this Godwin-Galton-Greg-Wallace-Fisher criticism? No, and hardly anyone else has either. It has remained a forgotten by-path in the history of Darwinian biology, and is now known only to a few persons of antiquarian bent.

The response of Darwin himself to the criticism was entirely and depressingly characteristic. He discusses at length the relevant writings of Greg, Wallace, and Galton, in chapter V of *The Descent of Man* (1871). Yet he somehow manages to do so without ever once betraying the faintest awareness that what he is dealing with is an *objection to* his theory. Well, that was Darwin's way. He was temperamentally allergic to controversy, and would always, if he could, either ignore or else candidly expound a criticism of his theory, as a substitute for answering it. The result might be, and often was, that his own position became hopelessly unclear, or else clear but inconsistent. But then, he did not mind *that* at all!

Greg and Wallace must have felt utterly baffled by this policy of masterly inaction, "confusionismo," and passive resistance on Darwin's part. Wallace had certainly succeeded in letting *other* people know that he disagreed with Darwin about natural selection and man; but it appeared quite impossible for him to let *Darwin* know. And Greg, for his part, had done everything that he could to let Darwin know it was a *criticism* which he was advancing. For example, he had entitled his 1868 article, to which Darwin refers without giving the title, "The Failure of 'Natural Selection' in the Case of Man." How could he possibly have written more plainly than that? But it was all in vain. Criticizing Darwin was like punching a feather mattress.

With Galton and Fisher, the situation is even more puzzling and amazing. For neither of these writers gives any sign of realizing, *himself*, that what he is saying is a *refutation of* Darwin and Malthus. Yet one would have thought it sufficiently obvious that

Galton's self-extinguishing peers, exempt from misery and un-tempted by vice, should by the theory have been not "disastrous," but *impossible*. And equally obvious that, more generally, Fisher's "social promotion of infertility," if a fact, *proves the falsity* of the Malthusian theory of population, on which Darwin's theory of natural selection rested. But, somehow, it was *not* obvious, either to Galton, or to Fisher, or to most of their readers.

I cannot explain this, at any rate beyond reminding the reader of something I said in Essay I above: that Darwinian Hard Men in general, and the eugenists among them in particular—to which class Fisher as well as Galton belonged—are *constitutionally* confused as to whether unDarwinian aspects of human life are injurious, or impossible. Neither do I see how it will ever be pos-sible, now, for the line of criticism I have been speaking of to receive the attention it deserves. If R. A. Fisher's powerful intel-ligence, in what is by general consent the most seminal biological book of the twentieth century, could not make clear even to him-self, let alone to his readers, that if he was right about man then Malthus and Darwin were wrong, it seems entirely out of the question that anyone else should be able to do so.

MALTHUS SAID in the first edition of his *Essay* that human num-bers are prevented from increasing at an extraordinary rate—say, doubling in every twenty years or less—principally by food shortage, war, and disease, or by one form or another of sexual vice. It is instructive to imagine this same proposition being ex-pressed in different words, and being published, as Malthus's version of it was, to the world.

It could be expressed as follows:

> If at the end of your life it is found that you had fewer children than can be explained by reference to the contingencies of food shortage, war, and disease to which you were exposed, the only explanation worth considering is that you engaged more or less in some form of sexual immorality such as infanticide, abortion, contraception or homosexuality.

Suppose, for a moment, that one entirely agrees with Malthus as to the immorality of infanticide, abortion, contraception, and homosexuality. Even so, would not publication of the statement above constitute some sort of record for insolence?

Think of three people whose reproductive careers, we will suppose, were never affected by food shortage, war, or disease. One is a nun of exemplary character who dies a virgin in fulfillment of her religious vows; another is a suburban housewife who only ever had one child, because she and her husband could not afford more; the third is a Don Juan who dies childless only because of his unflagging attention to contraception, abortion, and infanticide. Since we have excluded by supposition the influence of "misery" in explaining these low reproductive careers, we can only, according to Malthus, ascribe them to "vice." He *has* no third category.

There is something almost heroic about an insult as vast and undiscriminating as this. And yet similar things are by no means unknown. We would not need to look far into the literature of neo-Darwinism at the present day to find insults to the human race which would bear comparison with Malthus's. The reason is obvious enough, too. Namely, that a biologist, or anyone who looks at human life exclusively from the biological point of view, is peculiarly likely to blunder into just this kind and scale of insolence.

But suppose—what is more likely—that we do *not* share Malthus's moral conviction about abortion, contraception, and the rest. Put aside all terms of disapprobation, such as "vice" and "immorality." Let us simply *list* abortion, contraception, etc., as things which do, as a matter of fact, tend to repress human increase, independently of war, pestilence, and famine.

Now, Malthus's proposition will again constitute some kind of record, not for insolence this time, but for glaring falsity. For it contains not the faintest hint of recognition that, among humans, such things as pride, prejudice, and prudence are ever among the checks to population. Not one word of acknowledgment of the inexhaustibly many ways in which human beings differ from one another, in respects likely to influence their reproductive career: in interests, abilities, character, tastes, intelligence, information, beliefs, upbringing, circumstances . . . ! Nothing but the blank biological fact which is common to us all, and common to pines and flies: that the tendency to increase is checked by food shortage, disease, and large-scale killing. That, plus infanticide and a few other accomplishments peculiar to our species, make up Malthus's whole account of the checks to human increase. It would be worth a good deal to know what his contemporary, Jane Austen, thought of the adequacy of this account.

This was occupying high ground with a vengeance, ground conspicuous for insolence and falsity alike. But Malthus beat a precipitate retreat from this exposed position in the second edition of his *Essay*, published in 1803. Misery and vice, he now says, are not after all the only checks worth mentioning to human increase. There is another one, which he calls "moral restraint." And this new position is the one which he continued to occupy through all the later editions of the book.

With Malthus, the phrase "moral restraint" does not mean anything like as much as one might expect it to. He explains carefully what he does mean by it. "By moral restraint I would be understood to mean a restraint from marriage from prudential motives, with a conduct strictly moral during the period of the restraint . . . ."[16] That is, sexual intercourse being refrained from before marriage, and marriage being postponed until the economic means exist of supporting the children which marriage can be expected to produce.

It may not be obvious at once how very great a retreat Malthus made, when he admitted the existence of moral restraint as a check to population additional to misery and vice. But he did not merely acknowledge its existence. From the second edition of the *Essay* on, this newly discovered check occupied very many pages, including whole chapters, of the book. Admittedly, most of these pages were devoted to *recommending* this check, as being preferable to both the misery check and to the vice check. But Malthus also makes ample acknowledgment of moral restraint as a check to population which is *actually and importantly in operation.* He also says emphatically that mere *prudence* is a powerful check to population at the time he is writing, even if the other part of his definition of "moral restraint"—the part about "strictly moral conduct" while marriage is being deferred—is seldom satisfied in fact.

Thus, for example, he writes that although this virtue, of strictly moral conduct pending marriage,

> does not at present prevail much among the male parts of society, yet I am strongly disposed to believe that it prevails more than in [less civilized or earlier states]; and it can scarcely be doubted that in modern Europe a much larger proportion of women pass a considerable part of their lives in the exercise of this virtue than in past times and among uncivilized nations. But however that may be, if we consider only the general term ["moral restraint"] which implies principally a delay of the marriage union from prudential considerations, . . . it may be considered . . . as the most powerful of the checks which in modern Europe keep down the population to the level of the means of subsistence.[17]

How perfectly extraordinary! Prudence—a thing which in the first edition had not so much as merited a mention as a check to population alongside misery and vice—is now found to be, not only another such check, but, at least in modern Europe, *the most powerful* of all such checks! If it is, how could Malthus possibly have overlooked it when he was first writing the book five years before? And if he was right to recognize it in 1803—as he plainly was—then he must have been profoundly wrong to omit it in 1798.

By making this retreat, Malthus tacitly gave up the premise of his biological argument against Godwin and other communists or socialists. But this was, if anything, a gain rather than a loss; because that argument had always been silly anyway.

The argument had been, in essence, that if private property were abolished, population would press upon the means of subsistence. Yet the very premise of the argument was that population *already does* press upon the means of subsistence, and does so always and everywhere, in every species of organisms. Malthus was therefore in the position of a worried parent who said to a fractious child, "Stop that, or we will all be breathing air!" or, "If you keep doing that, the sun will rise in the east." You cannot intelligibly threaten or warn a man that he is tending to bring about a deplorable state of affairs, while also saying that that state of affairs always exists, whatever his or anyone's conduct may be. (This absurd aspect of Malthus's argument was pointed out, though not very distinctly, by Godwin himself.)[18]

Malthus had other and far better arguments against "systems of equality": economic as distinct from biological arguments, and these are unaffected by his retreat. His economic argument against communism was that where no one could hope to improve their own or their children's economic position, and no one need fear to worsen it, no one would have a sufficient motive to work or save or limit the number of their children. His main argument against "creeping socialism," such as the Poor Law system, was that it *created* the poverty it was intended to relieve: both by economically rewarding those who depend upon the system, and by economically penalizing those who do not. Neither of these arguments, it will be obvious, rests at all on the assumption that population always presses upon the food supply, or on any assumption of anything like that.

What is important about Malthus's retreat is that it was an admission, not that the economics of his first edition had been wrong, but that its *biology* had. Wrong, that is, in assimilating the relation between population and food in the human case, to the case of pines, cod, and flies. You cannot consistently say that prudence is a powerful check to increase in the human case, and also say that, in humans as in all other species, population is always as great as the food supply allows; or even that it is always as great as is allowed by the food supply plus the prevalence of vice. Malthus's retreat from this blanket biologism, which he had at first embraced, was to the credit of both his common sense and his character. But it was also fatal to his biology, as well as to his consistency.

It may nevertheless still be true (as I have already said in earlier essays) that Malthus's principle *does* hold good for all non-human species, or nearly enough hold good, to make that principle a vital clue to the understanding of their evolution. I believe, indeed, that this is the case. If it is, Darwin and Wallace may have been prompted by a sound instinct, when they took from Malthus's book an *unqualified* principle of population: one which did *not* make an exception of man. All the same, it is ironic that they took this principle from one of the editions of Malthus's book in which the author himself had, very publicly, given it up. *He* had come round to admitting that our species is very different from all the rest, even if Darwin and Wallace had not.

It is sometimes believed that Malthus's retreat was forced upon him, by a criticism which his main intended victim, Godwin, had published in 1801. This could easily be true. Godwin's criticism was so short, true, and fatal, that it could easily have brought home to Malthus the insolence and falsity of what he had published in 1798. Godwin simply said that many people are restrained from marriage and parenthood neither by misery nor by vice, but by things like "virtue, pride, or prudence."[19] What a cool current of common sense and human nature, let into the hothouse biological atmosphere of Malthus's first edition!

But I have seen no convincing evidence that his retreat *was* forced by Godwin's criticism; and the supposition is unnecessary. For there are places in the first edition itself at which Malthus both recognizes the existence of "restraint from marriage from prudential motives," and recommends its wider practice. He says on pages 64–5 of that edition, for example, that such motives already restrain from marriage "a great number" of persons "of liberal education," and he goes on, on pages 66–9, to recommend this practice to persons who are not in that "rank of life." True, he later says (page 101), with glaring inconsistency, that all the checks to population "may be fairly resolved into misery and vice." But this is only to say that Malthus's inconsistence was not only between his first and all later editions, but present in the first edition itself.

The likeliest thing, it seems to me, is that it was Malthus's own common sense which compelled his retreat from the purely biological position which he had taken up at first. After all, pines and flies *cannot* refrain from reproduction on prudential grounds, and it is pointless to advise them to do so; but humans are different. And these facts are so extremely obvious that no one can forget them or by implication deny them. At least, no one can do so for very long; and Malthus had had five years in which to let his common sense get the better of his bad biology. We see here, incidentally, the superior literary strategy of Darwin. He avoided the necessity of introducing an embarrassing qualification about man into the second edition of the *Origin*, by the simple expedient of saying nothing at all about man in the first.

Essay 5
A Horse in the Bathroom, or
The Struggle for Life

... of the many individuals *of any species* which are periodically born, but a small number can survive.

—Darwin, *The Origin of Species*

I HAVE NO DIFFICULTY in accepting the *fact* of evolution. The proposition, for example, that existing species have all evolved from others, is not at odds with any rational belief that I know of. But I do not believe the Darwinian *theory* or explanation of evolution. There are several reasons. One of them is, that if that theory were true, then a struggle for life would always be going on among the members of every species; where in our species at any rate, no such struggle is observable.

This is a very obvious objection, of course, and would suggest itself to even the dullest person, once the Darwinian theory had been put before him. It is so obvious, indeed, that it always embarrasses me to put it to Darwinians, and embarrasses them to have it put to them. People as clever as Darwin and Wallace, therefore, must have seen this objection coming a mile off, and presumably each of them had some reply to it which satisfied his own mind at least. But if they did, they never told the public what these replies were. Darwin not only never replied in print to this obvious objection: he never directly adverted to it at all. The same is true of Wallace, as far as I know: though there are a few of his published works which I have not read.

But both men did publish what they clearly intended as a reply, though an indirect one, to the objection that there is no observable struggle for life in the human species. It was exactly the same reply from both of them. And they both put it in exactly the same place: namely, at the start of their respective chapters, in their main books, on the struggle for life. (These were Chapter III of Darwin's *The Origin of Species*, and Chapter II of Wallace's *Darwinism*.) Their reply goes as follows, except that, unlike them, I have made its purpose explicit in the first clause of my paraphrase.

> The absence of an observable struggle for life in the case of humans is not a serious objection to the Darwinian theory, because the struggle for life is not in general observable in animals or plants either, and yet we know it is going on all the time. For we know that every year each adult pine releases many thousands of viable seeds into the air, every pair of adult cod launches a million fertilized eggs into the ocean, and so on; and yet we also know that, by and large, the numbers of pines or cod is little or no greater at the end of each year than it was at the start. The vast majority of pine seeds and seedlings, and of cod eggs and young cod, therefore *must* have died in the course of the year. And it is essentially the same with every other species of organisms. Malthus' principle holds good in all of them alike. Population is always pressing on the supply of food available, and tending to increase beyond it. There must therefore be, in every species, a constant and universal struggle for life: a competition so severe that every year the great majority of young competitors *must* die. And yet we *see* nothing of this perpetual and universal destruction of young life; and we are therefore apt to conclude—mistakenly—that it does not exist.

Here is Wallace, setting out on this reply. "To most persons, nature appears calm, orderly, and peaceful. They see the birds singing in the trees, the insects hovering over the flowers, . . . and all living things in the possession of health and vigour, and in the enjoyment of a sunny existence. But they do not see, and hardly ever think of, the means by which this beauty and harmony and enjoyment is brought about. They do not see [for example] the constant and daily search after food, the failure to obtain which means weakness or death . . . ,"[1] etc., etc. Likewise Darwin begins his reply thus: "We behold the face of nature bright with gladness, we often see superabundance of food; we do not see, or we forget,"[2] the opposite of the coin. In short, we generally see only the winners in the struggle for life, and *they*, unsurprisingly, are healthy and happy; the losers are not, but we do not see *them*.

Almost everything that could be wrong *is* wrong with this reply of Darwin and Wallace. First, as to its method. It is an unsatisfactory way of defending a scientific theory, when it is objected that what it predicts is not observable in one area, to reply that that is not a problem, because what it predicts is not observable anywhere else either.

Then, as to fact. The Darwin-Wallace reply has the implication that child mortality is about the same in all species, or at least is tremendously high in all. "Child mortality" is just the proportion of individuals born which die before reaching reproductive age. In many species, of course, including cod and pines, it *is* enormously high: 99 percent or even more, according to competent authorities, in the case of cod for example. But surely no sane person will believe that child mortality is anything like as high as that *across the board*: in all birds, all mammals, all everything? A female elephant (Darwin tells us) has six offspring in a lifetime:[3] so how would elephants get on under a child mortality of 90 percent or so, or anywhere near that?

Further, the reason that Darwin and Wallace give, why we do not constantly *see* the slaughter of the innocents in nature, is ridiculous. The real reasons have nothing to do with the face of nature "appearing bright with gladness." The reason we don't see the killing of billions of young cod each year is simply that it takes place below the surface of the ocean. The reason we don't see millions of pine seeds or seedlings die each year is just that they are very small, completely uninteresting to most people, and anyway are most numerous where people are few. Why don't we see a lot more young birds lying around dead than we do? For several reasons, but nature's appearing "calm, orderly, and peaceful" is not among them. One reason is that birds—unlike insects, which do not care either way—dislike humans observing them, and young birds who are starving, or sick, or injured, dislike it even more than most. A second reason is that in the animal world the dead meat removal industry is big and efficient. A third is that the predators which kill young birds nearly always *eat* the bleeding things, and feathers soon blow away.

Wallace's "nature appearing calm, orderly, and peaceful," Darwin's "face of nature appearing bright with gladness": these flowers of Victorian literature are not of a kind that lasts well, but of course they had a serious purpose at the time. It was to insinuate that anyone who does not admit that a struggle for life is always going on in *all* nature, human or non-human, is unable or unwilling to face the hideous scientific facts which Darwin and Wallace are trying to impart. Of course there is some truth in this insinuation; but not much. However able and willing you are to face the facts, you cannot get to see the struggle for life among ticks, earthworms, arctic lichens, the characteristic bacteria of the human digestive system, or a million other species. At least you cannot unless you have many millions of dollars, and much time, to devote to this purpose. And you could very easily fail to see it, even after all that.

In fact the face of nature, far from appearing to us glad when it does not, does not appear glad when it is. When are animals glad? During copulation, one supposes. At any rate it is hard to think of a likelier suggestion. But just as we do not see the one-billionth part of the deaths that conspecifics inflict on one another, neither do we see the one-billionth part of the gladness that they afford one another in copulation. And the little copulation we do see does not *appear* glad, or calm, orderly and peaceful, either. Rather the reverse, in fact, at least among large mammals, where copulation looks more like an unpleasantly one-sided fight.

Anyway, we *can* see the struggle for life among animals and plants, or at least quite a lot of instances of it; and we do see them, easily and often. Do you want to see hungry carnivores competing fiercely for even the smallest scrap of even the worst meat? Then you need do no more than go fishing on any beach, throw a bit of your smelly bait onto the sand, and watch the gulls. If you watch the same flock of gulls for a few days, it is quite likely you will learn to identify a particular gull who is a "born loser" and hardly gets any of the food; or to identify another gull who is "the boy most likely" (or the girl). A small suburban child who keeps rabbits in a wood and wire hutch will sometimes see, when the hutch is moved onto a new patch of grass, some of these not-very-competitive creatures "shouldering" one another, in order to get at the best grass, or more of it. Even the dullest witted gardener sometimes sees that three weeds of the same kind, which are competing for life with one of his precious plants, are also competing with each other; and also sees that one of the weeds is leaving the other two further behind every day. After such commonplace observations, natural and reasonable inductive inference does the rest: we conclude that the losing gull and the losing weeds will probably go on losing, and will die sooner, and leave fewer offspring, than their fellows who are better equipped to succeed in the struggle for life.

In animals and plants, then, the struggle for life *is* observable. Or observable enough, anyway, to satisfy anyone except a philosophical skeptic. Well, it had better be! If there were no observational evidence at all for the Darwinian theory of the origin of species, why would it be preferable to the "creation" theory? There is no observational evidence for that, either. But in fact the struggle for life *is* sometimes observable, and Darwin and Wallace themselves, in the very pages in which they are saying that it is not *in general* observable among animals and plants, are careful to avoid saying that it *never* is observable.

But now, where is it observable in *human* life? Where can one see even little bits of the struggle for life among our conspecifics? I mean, in our day to day life. It may be true that the survivors of a ship-wreck plane crash sometimes compete with one another for the last mouthful of water. Episodes of this kind nowadays exercise a fascination over people's minds, because, of course, we are all Darwinians now. But they really have nothing to do with the case. The Darwinian theory says that the members of every species are always and everywhere engaged in a struggle for life with one another. And what I say is, where can one see some instances of this struggle in *Homo sapiens*? I can easily tell Darwinians—in fact I just have told them—where they can see something of the struggle for life among gulls, rabbits, and weeds. All I ask is that some Darwinian should return the courtesy and tell me where I can see something of the struggle in *human* life. If the struggle is constant and universal, it ought to be the easiest thing in the world to point out instances of it. So please, would some Darwinians tell me where to find some?

But I hope no one will tell me what the locals always tell the visiting fisherman: "Ah mate, you should have been here *last* week." Will tell me, I mean, the stupid story which T. H. Huxley and many other Darwinians tell, about the old days: the story that, when people lived in "nature" or in "the state of nature" or in "the savage state," "life was a continual free fight, and . . . the Hobbesian war of each against all was the normal state of existence."[4] This is what I have elsewhere called "the Cave Man way out" of Darwinism's dilemma about human life. It is not even a biologically possible story, since "a continual free fight" between an adult and an infant, or a man and a woman, could not be of long duration or uncertain outcome. But since I have discussed this Darwinian fairytale elsewhere, I do not intend to go over it again here.

All Darwinians have a remarkable asymmetry of mind where their own species is concerned. On the one hand there is the human life which, both by experience and by reading history and literature, they know a great deal about; but all of this they put to one side, as having nothing to do with theory. They *have* to put it aside, because of course *this* human life contains not a single instance of the famous Darwinian struggle, and in fact consists entirely of *dis*confirmations of that theory. But on the other hand, Darwinians draw endless *con*firmations of their theory from the lives of extinct or hypothetical or imaginary or impossible human beings concerning whom they know exactly as much as the rest of us do: namely nothing. For Darwinians, where their own species is concerned, it's not what you know that counts: it's what you don't know.

Even more embarrassing than the Huxleian-Hobbesian story is what some other Darwinian friends say to me, when I ask them for instances of the struggle for life among humans. Namely: "You would have to admit that you have led an unusually sheltered life. If you were to try the world of business, as I have, you would find that it is Darwinian enough. It's a jungle out there."

I would, beyond question, find the world of business shocking and disgusting. But if on the one hand I have the disadvantage of lacking experience of the business world, my studious habits have afforded me the advantage, on the other hand, of some knowledge of Darwinism. Enough to know, at least, that even if the world of business is some sort of jungle, it is quite certainly not a *Darwinian* jungle. In the business world, every single transaction depends upon who owns what, and how much, money, shares, goods, real estate, or whatever. It depends, in other words, upon there being *right of property*. But there cannot be rights of property in a Darwinian jungle! Unless, indeed, there can be policeman pines, part of whose job it is to protect pine property, cod courts which determine which fish owns what, and bandicoot laws which regulate bankruptcy.

EVERYONE KNOWS that mammals and birds never have very many offspring at once, but take a considerable amount of trouble over the ones they do have, whereas pines and cod and many other organisms take little or no trouble over their offspring, but trust to "the law of large numbers" for carrying on their kind. So, when competent biologists tell us, for example, that 90 percent of young cod never get near reproductive age but die in their first four weeks of life, or that 98 percent of viable pine seeds will never live to produce seeds themselves, we have no difficulty in believing their statements.

But suppose you met a man who told you that child mortality is roughly of that same order of magnitude *in all species whatever*. To put a figure on it, suppose he said: "In all species, at least 90 percent die before reaching reproductive age." Then you would *not* believe him. It would not matter who he was: the most famous living biologist, or whatever. In fact you would think he must be mad.

You would be right, too. There are many species of mammals, of birds, and of insects, in which the female produces at most one offspring per year. But no sane person will believe that in all these species *only one offspring* in ten, on the average, is successfully raised to reproductive age. That would mean only one, on the average, of all those born in ten years. Locally and temporarily, no doubt, child mortality in our species, or in any other, can rise as high as 90 percent or more, or even 100 percent, because of some exceptional famine, or disease, or accident of climatic or geological change. But a child mortality of 90 percent or more, as a characteristic of *species*, and of *all* species, is simply too ridiculous a proposition to be seriously entertained. Even for elephants, which have a gestation period of more than twenty months, it is not as hard as *that* to have a grandchild.

Even if this man were to back down by 10 percent, and say, "Oh well, at least 80 percent die before reproductive age in all species," it would improve matters extremely little. You would still think he must be mad, and you would still be right to think so. It is not at all uncommon for a human female to have as many as eight children who reach puberty. Even now, and even in Australia, there are hundreds of such families, perhaps thousands. They are not as common as they once were, and not as common as they now are in some other places; still, they are common enough. But on the hypothesis that human child mortality is 80 percent or more—neglecting multiple births, which make up only 1 percent of all human births—a woman who raised eight of her offspring to puberty would have to have, on the average, forty children at least. That, however, is almost certainly a *physiological* impossibility. It is, quite certainly, *demographically* unheard of. The absolute record for single human births is about thirty-two.

Even 75 percent or more, as a rate of child mortality in all species, is quite impossible. At that rate, a woman who got nine of her children to puberty would have to perform the impossible feat of having, on the average, thirty-six children all told. Yet many women have raised nine of their children to puberty.

Even 70 percent, as the species rate of child mortality in humans, is entirely out of the question. It would require a woman who got nine of her children to puberty to have, on the average, thirty children. But having thrity children is an absolute demographic prodigy, and has been so ever since demography existed: that is, roughly, for the last three and a half centuries. But women who raise nine of their children to puberty have been quite common during all that time.

The last few paragraphs will perhaps appear pointless. All right, child mortality cannot possibly be as high as (say) 80 percent in all species, because it cannot be as high as that in ours. But then, has anyone ever said or implied that it is? Anyone, at least, whom there is any need to take seriously?

Yes. Darwin implied that it is as high as that. So did Wallace. What is more, both men had accepted a certain theory which *requires* child mortality to be extremely high, in our species and in every other.

Darwin said, for example, that "each species, even where it most abounds, is constantly suffering enormous destruction at some period of life, from enemies or from competitors for the same place or food."[5] That every species, "during some season of the year, has to struggle for life, and to suffer great destruction."[6] He also says that the period of greatest destruction of animals and plants "in the great majority of cases is an early one":[7] that is, is destruction of *young* animals and young plants, and of eggs and seeds. That "many more individuals of each species are born than can possibly survive."[8] That "of the many individuals of any species which are periodically born, but a small number can survive."[9] And so on. And of course "surviving," among biologists, simply means that the organism survives to the age at which it is capable of reproducing in its turn.

Well, *of course*, Darwin said things like these! For they are, as everyone will recognize at once, essential elements of his explanation of evolution. Familiar as these statements are, however, it is worthwhile to consider how much they commit Darwin to, and how much anyone who accepts them is committed to.

It is perfectly clear, first, that Darwin was speaking about child mortality *in all species indifferently*. His own phrases quoted above—"each species," "of any species," "of each species"—settle that point beyond dispute, and leave no room for him to make any exception whatever to his generalization.

It is equally clear, second, that Darwin was saying *the same thing* about child mortality in all species. Namely, that it is—well, what shall we say?—"enormously high," or "tremendously high," or "extremely high," in all species alike. Any of those three phrases would do, to express what he clearly intends.

Let us settle for "extremely high." So Darwin said that child mortality is extremely high in all species. But how high is extremely high? Or, going back to his own phrases, when he said that in every species many more are born than can survive, *how many* more is "many more"? And when he said that in each species "but a small number" of those born can survive, what sort of percentage did he mean by "but a small number" (that is, "only a small number")?

It is not beyond the power of common sense to give an answer to these questions which is definite enough to be interesting, and therefore to tell us, nearly enough, what it was that Darwin was saying about the rate of child mortality in all species alike. Some people will, of course, be scandalized at the idea that common sense has any jurisdiction at all in such a textual matter. Yet it is only common sense, after all, which can assure us that Darwin did *not* mean, for example, that child mortality is 99.9 percent in all species; or that it is 100 percent, for that matter. So we can hardly afford to dispense with whatever help common sense can give us.

If there were, out of a hundred, sixty non-survivors and forty survivors, would anyone, either in 1859 or now, call this "many more" non-survivors than survivors? I say, absolutely not. It would not matter what the subject or the context might be: child mortality, or the aftermath of an aeroplane crash, or whatever. Even sixty-five non-survivors to thirty-five survivors would never be described, by any responsible writer, as many more non-survivors than survivors. Out of a hundred, seventy non-survivors to thirty survivors is the very least that could be properly described as "many more" non-survivors than survivors.

Then, as to "only a small number" of those born surviving: what is *only a small number*, out of a hundred? It could not possibly be as many as thirty, obviously. It could not even be, say, twenty-three. Suppose you believed, on the authority of someone or other, that out of a hundred only a small number survived, and it turned out later that in fact twenty-three had survived. Then you would believe, and rationally believe, that your authority had either made a mistake, or else had a great deal of explaining to do. In fact you would rationally believe that your authority was mistaken, or else had plenty of explaining to do, if it turned out that even 19 had survived. In other words, "only a small number," out of a hundred, *cannot* be as many as twenty.

I therefore say that Darwin, when he said that in every species many more are born than can survive, meant that child mortality is at least 70 percent; and that when he said that in each species only a small number of those born can survive, he meant that child mortality is 80 percent at least.

It is possible, of course, that my interpretations of Darwin's phrases are wrong, or forced. But even if they are wrong, can they possibly be *far* wrong? We know that he meant that child mortality is extremely high in all species. And then, there are his repeated references to *great* destruction, *enormous* destruction, in all species, concentrated, in the great majority of species, in the early period of life, with *only a small number* of those born surviving. If all of this does not add up to a child mortality of 80 percent at least in all species, it *must* mean some percentage only a little lower than that. Otherwise, we can all simply give up. The entire *Origin of Species* might just as well be a code message for something else altogether: for "A merry Christmas to all our readers," say, or for what you will.

That child mortality is, in all species, nearly 80 percent at least, is an incredible proposition. Yet it is, we see, implied by Darwin in his chapter on the struggle for life.

There is another thing about that chapter which is almost equally staggering. Namely, that it maintains a virtually complete silence on the *differences* in child mortality between some species and others. That is, Darwin was not content with implying that child mortality is extremely high in all species alike. He only once mentions the fact that species differ *at all* in child mortality; and he *never* mentions the fact that child mortality is enormously higher in some species than in others.

I cannot substantiate this part of my case, of course, by quotations. One cannot prove that an author is silent on a certain subject, by quoting his silence. Still, anyone who takes the trouble to read Chapter III of the *Origin* will be able easily to verify what I have just said. Darwin does not, of course, *say* that there are no differences, or only small differences, between different species in child mortality. But neither does he say that there are huge differences in certain cases, or even say, except once, that there are any differences at all in the child mortalities of different species. He just does not say! If this assertion appears incredible to the reader, then he need only read the third chapter of the *Origin* in order to satisfy his mind as to the fact.

While the reader is at it, he might as well also read the corresponding chapter of Wallace's *Darwinism*: Chapter II, "The Struggle for Existence." He will find that exactly the same improbable silence, on the subject of differences between species in child mortality, is maintained there too. And by that time, the reader may well begin to believe, as I do, that Darwin and Wallace intentionally concealed from their readers (whether consciously or not) a glaring and grave defect of their theory of evolution.

A reader of Darwin's Chapter III is likely to think that differences between species in child mortality are about to be mentioned, when he comes to the part about what Darwin calls "slow breeders." About, that is, "slow-breeding man"[10] (as he calls us), elephants, horses, etc., as well as certain birds and insects which produce at most one egg per year;[11] as contrasted with the many species of animals and plants which produce many offspring each year. But the reader is doomed to disappointment. For Darwin says not one word about the child mortality of his "slow breeders," as compared with that of other species. In fact the only use he makes of them is to illustrate yet again the old and entirely general Malthusian point: that even humans and elephants, if their increase were unchecked, would in no very long time multiply so greatly that (as he says) "the world would not hold them."[12] All perfectly true, no doubt; but it implies nothing at all about their rate of child mortality, compared with that of other species.

Even Darwin's expression "slow-breeding man" seems to me an ill-chosen one, or at least superficial. Humans, elephants, and the like, could more properly be called *fast* breeders, on account of the great deal of parental care which they exercise. Pines and cod, by contrast, are if anything more entitled to be called slow breeders. They belong to the same school of thought as the American army: "To hell with taking aim, just fire an *awful* lot of shots." But this policy can hardly be said to meet in general with any marked degree of success, at least if rapid increase in absolute numbers is the criterion of success. As a means to that end, the mere multiplicity of your offspring appears to count for a good deal less than taking much care of the offspring, few as they may be, that you do have.

That point was suggested by Darwin himself, in the one sentence in his third chapter where he *does* mention differences in child mortality between species. This sentence reads: "If an animal can in any way protect its own eggs or young, a small number may be produced, and yet the average stock be fully kept up; but if many eggs or young are destroyed, many must be produced, or the species will become extinct."[13] It's not much, as you can see. But when you have read that, you have read all there is in Chapter III about the differences between species in child mortality. And if you happen to be a reader who is troubled by the painfully obvious question, "How could humans, elephants, horses, etc., possibly make a living under a child mortality of 80 percent or more?"— well then, you might as well spare yourself the trouble of reading Darwin's chapter on the struggle for life, because there is nothing in it which will help you to answer the question.

Exactly the same goes, yet again, for Wallace's chapter on the struggle for life. He too mentions the so-called slow breeders, but then only uses them to illustrate once again the old and general Malthusian point. And he too leaves the reader mystified, as to how, under the immense child mortality which he ascribes to all species indifferently, these "slow breeders" could possibly survive at all.

HERE, THEN, is an amazing historical fact: that there has been, lying on the very surface of Darwin's famous theory of evolution for nearly 150 years, the incredible proposition that child mortality in humans is about 80 percent at least.

And here is a second fact which, though not in that class, is still astounding enough. Namely, that in more than forty years extensive reading in the literature of Darwinism and its critics, I have never come across a single allusion to the fact that the Darwinian theory does contain this incredible proposition.

Yet one would have thought that the proposition, that human child mortality is about 80 percent at least, could no more have escaped notice and comment, most of it unfavorable, than a horse in the bathroom. I cannot afford, however, to make too much of this remarkable blindness which has afflicted other people. For it was only about two years ago that I first began to notice this horse in Darwin's bathroom myself.

But it almost certainly has been noticed before, by authors whom I have not read. For there is nowadays a branch of neo-Darwinism which is concerned with "r/K" theory: that is, with the spectrum of reproductive strategies between that of pines (say) with their many offspring but no parental care, and that of man, with few offspring but maximum parental care. It is reasonable to suppose that some of the more historically minded of these scientists have at some time read *The Origin of Species*. If they have, their hair must have stood on end, when they found Darwin implying that the difference in child mortality between pines and man, or between any two species, is either non-existent or too small to be worth mentioning. For this reason I believe that at least some of these people must have noticed the horse before I did.

It cannot possibly have escaped the mind of a Darwin or a Wallace, either that many women do get eight of their children to puberty, or that, in order to do so under a child mortality of 80 percent, these women would have to perform the impossible feat of having on average forty children each. Yet both of these men said things in their books which imply that child mortality is near enough 80 percent or more in *every* species. How can they have come to believe such a glaring falsity?

Darwin, for example, certainly knew that his wife Emma got seven of her children to puberty, and also knew that this was about par for the course among people in their social situation at the time: certainly nothing like a freakishly successful feat of child raising. Yet in print he implied that any woman who did what his wife did would have to perform the impossible feat of giving birth, on the average, thirty-five times. So *why* did he imply in his book that child mortality is about 80 percent or higher in all species?

We know only too well, alas, how this question would be answered by fashionable Marxist-Kuhnian historians of Darwinism. Their answer would go like this. "Darwin often mistook transient features of contemporary capitalism for permanent biological realities; and then, child mortality *was* colossal in Britain at his time. Three of the Darwins' own ten children died before puberty, although they belonged to the most privileged 15 percent of the population. Children of the underprivileged worked and lived in atrocious conditions. Unbridled *laissez faire* capitalism . . . callous indifference of the ruling classes . . . etc., etc.)."

Well, it is perhaps just possible that Darwin *was* a much more intellectually blinkered man than the people who now write malignant and supercilious books about him. But it is *not* possible that the many women in Darwin's time who got eight of their children to puberty had, on the average, forty children each, or anywhere near so many. So child mortality was *not* 80 percent in Darwin's time, or anywhere near so high. No, not even in "darkest England," and not in any other place or time either. Except, indeed, some place and time which is equally unknown both to demography and to the physiology of human reproduction.

The real reason why Darwin and Wallace enormously overestimated the rate of child mortality in humans is quite obvious, and lies right under our noses. They did so under the compulsion of a *theory*. Both men had embraced, in order to explain evolution, the Malthus-Darwin principle of population: that population always presses on the supply on food, and tends to increase beyond it. And this principle *does* require child mortality to be terrifically high, in our species and in every other. To see this, consider a small and schematically described, but still sufficiently realistic, example of a human population.

As Li'l Abner once sang, "It's a typical day in Dogpatch, U.S.A." The date is 1 January, 5000 B.C. Population of Dogpatch: 1,000. It's a typical enough year, too. In fact they have had a whole string of typical years lately. There have been no spectacular strokes of luck. They haven't just moved into a new territory full of food and empty of competitors, nor have they made any recent breakthroughs in agricultural or hunting technique, or in pediatric or geriatric medicine. But as against that, there hasn't been a spectacular famine, plague or war for decades, either. So population in Dogpatch is what the Malthus-Darwin principle says it always is everywhere: at or rapidly approaching the maximum permitted by the available food. Hence there is, in Dogpatch demography, neither a baby bulge or a gray bulge. There could hardly be the latter anyway: they "do it hard" in Dogpatch, and almost no one lives past fifty. Still, they have been doing all right lately, and even though, farmer-like, they won't admit it, they *did* have a little bit of luck last year. In fact there is today enough food available to support a population of 1030.

Now, the Malthus-Darwin principle tells us that this ecological niche will be filled this year. It is a principle concerning all species of organisms indifferently, and says that where there is food for a possible person, cod, or pine, there is already, or else will be as soon as is reproductively possible, an actual person, cod or pine.

Then, death last year carried off seventy of the older citizens of Dogpatch. This must be the main source of any hope that this year's beginners have of surviving. As Malthus is always saying, it is always death which makes possible nearly all the births that survive.[14] A surplus of food, by contrast, is just an occasional lucky windfall. It *cannot* be more than that, if the Malthus-Darwin principle is true: for that principle says that pressure of population is always busy ensuring that there will soon be *no* surplus of food. Junior officers of the Royal Navy in the eighteenth century used to drink a toast that ran: "Here's to a sick season and a bloody war." Meaning, of course, here's to many and early opportunities for promotion. This is a toast which, if the Malthus-Darwin principle is true, the beginners in every species of organisms should most heartily drink to. Or they could adopt as their motto what Hugh Trevor-Roper once said, about unwitting friends of the unborn such as Hitler, Stalin, and Mao Tse-Tung: "while there's death there's hope."

Of the 500 females in Dogpatch, almost none (as we saw) are of post-reproductive age. There is no "bulge" of pre-reproductives, either. So females of reproductive age must make up the majority of the 500: say, there are 300 of them. Now, on the Malthus-Darwin principle, all of these 300 women are going to reproduce this year. They can, so they will; just like pines or cod.

This year, then, there are just 100 openings for beginners in Dogpatch: thirty of them contributed by the food surplus, and seventy by last year's deaths. But there are going to be 300 births. In short, two-thirds of those born this year are going to die soon. And this is "from the simple principle of population,"[15] as Malthus puts it. The actual mortality among those born this year will be higher than two-thirds, because the deaths among them which are due to disease, war, etc., have still to be taken into account. It will be further raised by the deaths this year of some children who were born last year or earlier, but are still below reproductive age.

Yet every assumption that we have made, apart from the Malthus-Darwin principle, was in fact distinctly *favorable* to the prospects of this year's beginners. For a start, there was food for thirty more people in a population of 1,000. That is an opportunity for a 3 percent increase in population in one year, without the necessity for *any* increase in the supply of food: which is something not to be sneezed at in any population's language, and which must be, if the Malthus-Darwin principle is true, a rare bit of luck. Then, an overall mortality of seventy in 1,000, or 7 percent, is a very heavy rate, at least by all *historical* standards. So there's a second piece of luck for the beginners: their elders are dying rather fast. Finally, it could easily have happened, and often must happen, that more than three out of five females are of reproductive age. And no opportunity for reproduction is ever neglected, according to the Malthus-Darwin principle.

The Dogpatch elders speak the truth, therefore, when they say that young people nowadays are exceptionally fortunate, though they may not know it. All the same, according to the Malthus-Darwin principle, at least two-thirds of all those born this year must soon die. That is how severe the Darwinian struggle for life is among humans, and how high child mortality must be, on the

Malthus-Darwin principle, even where circumstances are exceptionally fortunate. Where there is no luck going round at all—which *must* be the great majority of cases—child mortality must be much more than two-thirds. If there were no increase of food, for example (though no decrease either), child mortality in Dogpatch would be 230 out of 300: about 77 percent. And this is still from "the simple principle of population": the deaths among those born this year from disease, etc., have still to be added. So have the deaths this year of pre-reproductives who were born last year or earlier.

If the Malthus-Darwin principle is true, this is how it must be with human populations, always and everywhere; except, as Malthus says, during the first "peopling of new countries,"[16] or immediately after an exceptionally devastating famine, epidemic, or war. Except in those necessarily rare cases, human child mortality must *always* be at least two-thirds.

But *in fact*, of course, that statement is simply and obviously false. In the only period for which anything at all is known about human demography—the last three and a half centuries—child mortality has usually been a good deal lower than the Malthus-Darwin principle says it must always be; and in all advanced countries during the last hundred years, it has seldom, if ever, been as high as even 20 percent. In the last fifty years, of course, child mortality has almost ceased to exist in all advanced countries.

Yet the Malthus-Darwin principle says that, except on necessarily rare occasions of good luck, child mortality in humans must always be two-thirds at least. A defender of that principle could, of course, say that all the human demography so far known, and especially the demography of the last hundred years in advanced countries, has been one continued piece of rare luck. But at that rate, luck could go on for *any* length of time, and over an area of any extent; and we might as well say straight out, that no one knows or can know anything at all, about what the *real* rate of human child mortality is.

But the only *rational* conclusion which can be drawn, from the facts that we do know about child mortality, is that the Malthus-Darwin principle is simply not true of humans. It may be true, or near enough true, of pines, cod, etc., that they always multiply with maximum speed up to the limit of population which food permits, tend to increase beyond that limit, and always suffer consequent penalties in child mortality. But it is not true of humans, or even near the truth, that *they* always do the same.

The Malthus-Darwin principle, however, is a part, and the major part, of the Darwinian theory of evolution. (The only other part is the proposition that, in every species, there is always variation among individuals in heritable attributes.) So, since it is rational to conclude that the Malthus-Darwin principle is false, it is rational to conclude that the Darwinian theory of evolution is false.

But of course no Darwinian will ever admit that. You can point out to him that child mortality has not been anything like two-thirds, in fact has not been even 20 percent, in any advanced country for the last hundred years. But all you will achieve by doing so is to propel him into his Cave Man mode, and start him talking, yet again, about his favorite topic: the old days. He will tell you about the high rate of child mortality that existed among humans when they were in their "natural" state, or "under natural selection," or were primitive hunter-gatherers, exposed to the full severity of the Darwinian struggle for life.

He will never admit that the low *actual* child mortality in advanced countries is any evidence at all against his theory. No, but he will point to the *hypothetical* high child mortality among our remote ancestors, as evidence *for* his theory. In fact, of course, he knows as little as you do about what the child mortality among ancient hunter-gatherers really was. The difference between the two of you is, that he has a theory which says that it *must* have been high, and that that is good enough for him.

You can try reminding the Darwinian, if you like, that this theory of evolution is a proposition about *all* species of organisms, at all times and places; and that man is a species, that the last three centuries are times, and that advanced countries are places. But you will be wasting your breath. In fact you will get more sense out of an intelligent creationist, and as much out of a log of wood.

Yet the Malthus-Darwin principle not only implies a high child mortality for all species at all times: by a curious irony, it actually does so even more effortlessly than Malthus, at any rate, realized. For he believed that, except at "the first peopling of a territory" or just after a terrific mortality from some cause or other, we and all other organisms are born "into *a world already possessed.*"[17] Which *is* true, of course, of most species. A beginner human or pine or cod finds itself in the presence of many conspecifics of an earlier generation, who have not only taken possession of all the most desirable real estate, but are busy mopping up most of the food available. No wonder if, in almost all species, most of the beginners struggle for life for a little while, and then struggle no more.

But Malthus had forgotten, or did not know, that it is not so in all species. There are many species of insects, especially wasps, in which the entire parental generation is dead by the time that the new generation hatch out of their eggs. In all annual plants, likewise, the parent plants are all dead when their seeds germinate. All these organisms, therefore, are so far from being born into a world already possessed, that they are born into a world which is totally empty of conspecifics belonging to the parental generation. That was the novelist Samuel Butler's idea of paradise, and is also the ideal of many other permanent adolescents. But it is an ideal which is realized in fact by many insects and plants.

And yet now see how effortlessly a Darwinian struggle for life must ensure *even in these species*, if the Malthus-Darwin principle is true. All that is required is, first, that the parent generation should have been as numerous as its food permitted, and second, that the offspring generation be more numerous than the parental. Then (as long as the food supply does not increase), there *must* be a struggle for life among the young conspecifics, and a resulting child mortality. The Malthus-Darwin principle ensures that the first condition will usually be satisfied. And the second condition will *in fact* be satisfied, by every species that is not actually dicing with death by having zero or negative population growth. In other words, on the Malthus-Darwin principle, there will be a Darwinian struggle among young conspecifics, and a child mortality, even where they are *not* born into a world already possessed. And the struggle will be the more severe, of course, and the child mortality will be the higher, the more the offspring generation outnumbers the parental.

David Hume was certainly no naturalist, except in the philosophical sense of that word, yet even he knew the above fact about certain animals.[18] It is therefore likely that Malthus too would have known it, and in any case, he surely must have known of the existence of annual plants. He did, however, have a strong motive for forgetting facts of this kind. For *his* only interest in the principle of population was to draw from it political conclusions favorable to the possessing classes of his own time and place. From this point of view, any reminder that in some species there is no such thing as a possessing generation when the new one arrives, would have sounded a discordant note. It might even have helped, for example, to revive such socially dangerous old myths as the admirable communism of bees.

So FAR in these essays I have meant, by "the Malthus-Darwin principle," or by "Malthus's principle of population," the proposition that in every species population always presses on the supply of food, and tends to increase beyond it; or, alternatively, the proposition that organic populations are always as large as their food supply permits, or else are rapidly approaching that limit. Now, let us call either of those propositions, "the plain principle" of population.

I call it so to distinguish it from the more arresting proposition which Malthus himself maintained, and which was usually what he meant when he spoke of "the principle of population." This was, that the supply of food can at most increase arithmetically (as do, for example, the numbers 2, 4, 6, 8, 10 . . .), whereas population always tends to increase geometrically (as do, for example, the numbers 2, 4, 8, 16, 32 . . .). Let us call this "the mathematical principle" of population. Malthus thought that the plain principle is simply an obvious, and less interesting, consequence of the mathematical one.

It is this mathematical principle of population with which Malthus's name was indissolubly associated during this lifetime, and for decades after his death in 1834. By now, however, it has been virtually forgotten, while the plain principle, because of the explanatory power which Darwin detected in it, has become ever more important. But the mathematical one, with its once famous "double ratios," has had no defender, that I know of, for more than a hundred years.

The reason (I take it) is that the mathematical principle, from the very first, attracted a host of criticisms from intelligent people. The criticisms came not just from communists, like William Godwin, or socialists like Robert Owen, or radicals like William Hazlitt and Francis Place; they also came from respected and justly respected economists, such as Nassau Senior, and from respected and justly respected philosophers, such as Richard Whately. And though these criticisms were, naturally, not all of equal weight, there was not a single one of them, at least as far as my reading extends, which completely missed its mark. I will give what seem to me the two most important of these many criticisms.

Whately made a most illuminating distinction between two senses of such words as "tends" and "tendency." "By a 'tendency' towards a certain result is sometimes meant, 'the existence of a cause which, if *operating unimpeded*, would produce that result. . . . But sometimes . . . 'a tendency towards a certain result' is understood to mean 'the existence of such a state of things that that result *may be expected to take place.*'"[19] In other words, something has a tendency to F, in the first sense, if and only if it *would* F if left to itself; while in the second sense, something has a tendency to F if and only if it *usually does* F.

There plainly can be a tendency in the first sense, without there being a tendency in the second sense. According to Newton's first law of motion, for example, there is in all bodies a tendency to continue at rest or in uniform rectilinear motion: a tendency which will be realized in fact, unless the body is acted upon by some unbalanced external force. But it does not follow from Newton's first law, and nobody will imagine for a moment that it does follow, that bodies *usually do* continue at rest or in uniform rectilinear motion. In fact the existence of the tendency does not even logically require that any body *ever* does so.

Likewise, Malthus's mathematical principle says *inter alia* that population always tends to increase geometrically. Let us suppose that this is true, in the first sense of "tends." Then the principle says that population always *would* increase geometrically, if it were left to itself, or "operated unimpeded." But it by no means follows that population tends to increase geometrically, in the *second* sense of "tends." It by no means follows, in other words, that populations *usually do* increase geometrically. In fact it does not even follow that they ever do. And, as a plain matter of demographic history, they almost *never* do, at any rate for more than two generations.

Whately directed this criticism, not at Malthus's mathematical principle, but at a certain "hyper-mathematical" principle of population, maintained by some hyper-Malthusian whom Whately does not name, to the effect that the pressure of population on the food supply must always become *more and more* severe. But the relevance of his criticism to Malthus's own principle is obvious, and can scarcely have escaped Whately's mind, or the mind of Malthus himself. For it shows that the truth of the mathematical principle does not require that there should always be, or even that there should ever be, any *actual* pressure of population on food at all. It does not entail that any *actual* population ever increases geometrically, any more than Newton's first law entails that any actual body continues at rest or in uniform rectilinear motion. Indeed, it would be perfectly consistent with the mathematical principle, if no population ever increased at all, geometrically or otherwise.

The second criticism of the mathematical principle seems to me almost as good as Whately's. It is one which was never, perhaps, made as distinctly as it could have been, but it was made by William Hazlitt,[20] for one. It is essentially as follows.

The mathematical principle contrasts the rate at which *population* tends to increase, with the highest rate at which *food* can increase. But the contrast makes no sense, because food, or at any rate the food of animals, *consists* of organic populations. The food of humans, for example, consists of parts or products of populations of cattle, wheat, fruit trees, bees, milking cows, etc. So if population tends to increase geometrically in all species, then that is true not just of human populations, but of cattle and wheat populations too: that is, of *human food*. And contrariwise, if *in fact* something nearly always prevents that tendency from being realized in the cattle and wheat on which we feed, then presumably something does or might nearly always prevent it from being realized in human populations too. And if it does, where is there any of the pressure, alleged to be universal and constant, of population upon food?

Malthus pictured our populations as always pressing on our food, with huge increases in the former being prevented only by the limited amount of the latter available. Well, we *are* organic populations, and we do eat food. But we are not just populations: *we are also food*, to countless species of bacteria, insects, and (according to circumstances) other animals, while we are alive, and to countless other species of organisms when we are dead. Even plants, including wheat and fruit trees, are being fed all the time by human and other animal detritus, excreta, and bodies. Now bacteria, mosquitoes, wheat and innumerable other species to which we are food, all multiply a great deal faster than "slow breeding man." According to the mathematical principle, all the species to which we are food are constantly tending, as we are, to multiply beyond the food available. So if *we* are in constant danger of geometrically increasing out of house and home, why are we not equally menaced by geometrically increasing mountains of bacteria, mosquitoes, and wheat?

EITHER OF THESE two criticisms, if I had been Malthus and had asserted the mathematical principle in print, would have had me worried sick. But Malthus must have been made of different stuff. Neither these nor any other criticisms ever shook his faith in the mathematical principle. It is still there in his sixth and last edition of the *Essay* in 1826, as it was in the first edition of 1798. It is, perhaps, given a little less prominence in the later editions than in the first. But it is still there, explicitly stated and defended, in all of them alike.

Yet it is most unlikely that Malthus did not know of Whately's criticism, which was published in 1832, when Malthus was still in good health and intellectually active. Besides, part of Senior's criticism was essentially identical with that of Whately, and Malthus certainly did know of it, because he corresponded with Senior on the matter. (See the latter's *Two Lectures on Population*, 1829) We know he had read Godwin's criticisms of the mathematical principle in *Of Population* (1820), because Malthus (quite scandalously) was given that book to review. It is impossible to believe that he would not have read the essay on "Mr. Malthus" in Hazlitt's *The Spirit of the Age* (1825). And if he did, he would have there come across the criticism about food *being* population.

One can therefore only conclude that Malthus considered the two criticisms I have mentioned, and all the others, either mistaken or unimportant. And if he did, well, there were intelligent people decades after his death who agreed with him. For the mathematical principle was in 1859 incorporated by Darwin into *The Origin of Species*, without any sign of doubt or uneasiness. He refers no fewer than nine times in that book to "all plants and animals . . . tending to increase at a geometrical ratio,"[21] to "the geometrical tendency to increase," and so on: and on two of these occasions he refers to Malthus by name.[22] Thirty years later still, in *Darwinism*, Wallace likewise refers to "the rapid increase, in a geometrical ratio, of all the species of animals and plants."[23] I do not know whether either Darwin or Wallace ever heard of the objections which had been made to the mathematical principle long before. But if they did, they must have thought that there was nothing in them.

If they did think so, however, they were wrong; and wrong in exactly the same way as Whately had pointed out that Malthus was. The mathematical principle is either straightforwardly false or else predictively impotent. It is false if "tends to increase geometrically" means "usually does increase geometrically." But if it means only "would increase geometrically if left to itself," then the principle could be true without there being, always or usually, any *actual* pressure of population on food at all, and hence without there being any struggle for life among conspecifics or consequent child mortality. It would not even follow that anything of that kind *ever* happens.

Clearly, it was the *plain* principle of population which both Malthus and Darwin needed for their respective purposes. Malthus needed a constant and universal pressure of population on food, for his biological argument against communists and other egalitarians. Even with that pressure (as I pointed out near the end of the preceding essay), that argument is a failure; but without it, Malthus would have had no biological argument at all. More importantly, it is clearly the plain principle which Darwin and Wallace needed, for their explanation of evolution. For the plain principle, as we have seen, *does* ensure that there will be in every species a severe struggle for life among conspecifics, and a high child mortality. And that is the very thing, of course, which is needed to ensure in turn that in every species there will be that natural selection which is, according to the Darwinian theory, the *vera causa* of evolution.

But now, is the plain principle of population any less exposed than the mathematical one, to the two objections I have mentioned? I will not here pursue the Hazlitt objection, based on the fact that food, at least among animals, *is* population: not because this objection does not deserve pursuing—far from it—but because I find it so bewildering that I cannot pursue it effectively. But what about the Whately objection, or a variant of it? After all, the word "tends," in which Whately detected an ambiguity fatal to the mathematical principle, also occurs in the plain one: that population always presses on food, and *tends* to increase beyond it. Perhaps "tends" suffers from a fatal ambiguity here too?

Besides (as will be obvious to any philosophical eye), the idea of tendency is present from the outset, in speaking of population as *pressing* upon food. Nor can we escape the idea of tendency, by resorting to the alternative version of the plain principle: "organic populations are always as large as their food supply permits, or else are rapidly approaching that limit." For, under the transparent disguise of the word "approaching," the idea of tendency is plainly present here too.

Well: "population always presses on food, and tends to increase beyond it." If "tends to increase beyond it" means only, "would increase beyond it if left to itself," then the plain principle is predictively impotent, just as the mathematical one is under the corresponding sense of "tends to increase geometrically." On this interpretation, the plain principle does not imply the occurrence of any actual struggle for life among conspecifics, or any actual child mortality, at all; any more than Newton's first law implies the existence of any actual body which is at rest or in uniform rectilinear motion.

If it had *this* meaning, the plain principle would clearly be no use whatever to the Darwinian theory of evolution. For that theory requires an actual struggle for life among conspecifics, always and everywhere, and an actual child mortality: anything less would not be enough to bring about *actual* natural selection, always and everywhere. A merely hypothetical natural selection, which *would* occur *if* population were left to itself, would not suffice for Darwinism at all.

So let us try the other sense of "tends": the sense in which "tends to F" means "usually does F." Then the plain principle comes out this way: "population always presses on food, and *usually does* increase beyond it." And this, by contrast, is "just what the doctor ordered" for the Darwinian theory of evolution. For now the plain principle *does* imply that in every species there will be, at least usually, a struggle for life, a child mortality, and consequent natural selection of the individuals best fitted to survive.

We see, then, exactly what it is that the Darwinian theory requires, in the way of a Malthusian principle of population. First, it must be, not the mathematical principle, but the plain one: that population always presses on food, and tends to increase beyond it. And second, "tends" must be understood here, not in the "thin" sense, in which "tends to F" means only "would F if left to itself," but in the "thick" sense, in which it means "usually does F."

But even this, alas, is not *quite* what the doctor ordered for Darwinism. For according to the Darwinian theory, natural selection goes on always and everywhere in every species. Now, for natural selection to be going on, a struggle for life, with a child mortality, has to be going on. And for *that* to be going on, there has to be an excess of population above the number which there is food to support. So, if natural selection is going on always and everywhere, then there has to be such an excess, not just *usually*, but always. But this is to say that every species is always actually *increasing*, and hence need never fear extinction, or even declining or stationary numbers!

Think of a population which has increased up to its Malthusian limit—that is, has become exactly as numerous as there is food to support—and which, from that time on, proceeds to decline or at least not increase in numbers, while its food supply remains what it was at the limit. There will then be no struggle for life among conspecifics, and no child mortality arising from that struggle: for there will be enough food to support all that are born. There will, consequently, be no natural selection going on: "the examinations set down for today have been cancelled." Yet, according to the Darwinian theory, natural selection goes on always and everywhere. Whence it follows that there cannot be any such population as this one, with its declining or stationary numbers: all populations must always increase in numbers.

The Darwinian theory is in fact, then, exposed to a fatal objection which is essentially identical with the one made by Whately to Malthus's mathematical principle of population. If the word "tends" bears only its "thin" sense, then Darwin's theory and Malthus's principle alike fail to predict *any* pressure of population on food, any struggle for life, any child mortality, or any natural selection. If, on the other hand, "tends" bears its "thick" sense, then both Darwin's theory and Malthus's principle predict *too much* pressure of population on food, too much struggle for life, too much child mortality, and too much natural selection: that is, more than really exists.

"OH DARWINIAN STRUGGLE for life—what crimes have been committed in thy name!"

Some of these crimes are well known, and have even become a political cliché. Namely, those of the American capitalists who, late in the last century and early in the present one, used the Darwinian idea of a universal struggle for life as a scientific justification of their ruthless acquisitiveness.[24] It is less well known, but still is fairly well known, that Adolf Hitler found or thought he found an authorization for his policies in the Darwinian theory of evolution. He said, for example, that "if we did not respect the law of nature, imposing our will by the right of the stronger, a day would come when the wild animals would again devour us—then the insects would eat the wild animals, and finally nothing would exist except the microbes. By means of the struggle the elites are continually renewed. The law of selection justifies this incessant struggle by allowing the survival of the fittest. Christianity is a rebellion against natural law, a protest against nature."[25]

What deserves to be well known, but has in fact been virtually forgotten, is this: that if Darwinism once furnished a justification, retrospective or prospective, for the crimes of capitalists or National Socialists, it performed the same office to an even greater extent, between about 1880 and 1920, for the crimes, already committed or still to be committed, of Marxists. It is in fact scarcely possible to exaggerate the extent to which Marxist thought in this period incorporated Darwinism as an essential component. Marxists do not believe, of course, that there will be any struggle for life among human beings in the future classless society. But it was that Darwinian conception which Marxists at this time adopted as their description of human life *under capitalism*.

The reader can easily verify this statement, by opening any Marxist book, pamphlet, or newspaper of that period, whether written by an American Marxist, a Russian one, a German, or whatever. For example, an American book which borrowed its title from Darwin and Wallace, *The Struggle for Existence*, and which, despite being a very large volume, had reached its seventh edition by about 1904:[26] what sort of book would *that* have been? Hardly anyone nowadays could guess the right answer to this question. But to anyone familiar with the Marxist literature of this period, the right answer will be obvious: it was a manual of *Marxism*.

In Russia in the 1880s, numerous small groups contended with one anther for the leadership of the entire communist-terrorist movement. Sergius Stepniak was the leader of one of these groups, and he published a collection of his pamphlets, *Nihilism As It Is*, in about 1893. In this book he rests his own group's claim to the leadership on its having arisen, from other "incomplete organizations, by virtue of natural selection"[27] under Czarist pressure.

Of course Stepniak and W. T. Mills (who wrote *The Struggle for Existence*) are authors now forgotten. But not all the authors who combined Marxism with Darwinism have been forgotten. Jack London is one who has not. Another is Upton Sinclair, whose powerful Marxist novel *The Jungle* (1906) portrays life in Chicago under capitalism as life in a Darwinian jungle. Yet another is August Bebel, the leader of Marxist Social Democracy in Germany in the late nineteenth century. His main book—and a good book too—was *Woman Under Socialism* (1879),[28] which is a perfect example of the blending of Darwinism with Marxism, especially in Chapter V of its longest section, "Woman in the present." The 1904 English translation of this book, I may add, was from the *thirty-third* German edition: a fact which will indicate how far from being idiosyncratic was Bebel's combination of Darwinism with Marxism.

A present-day reader is likely to think that this combination, even if it was once common, must always have been an incongruous one. But that is a complete mistake, and can only arise from ignorance, or from a present-day Marxist hostility to "biologism." If the combination of Darwin with Marx were incongruous, then a book like Bebel's would have to be full of leaky joints or weak stitches, wherever the author's Marxism ended and his Darwinism began, or vice versa. But anyone who attempts to identify any such weak points in *Woman Under Socialism* will find his attempt more educational than successful.

It will perhaps be said, in defense of Darwinism, that many and enormous crimes have been committed in the name of *every* large and influential body of ideas bearing on human life. Whether that is true or not, I do not know. But even if it is, there are great and obvious differences, among such bodies of ideas, in how easily and naturally they amount to incitement to the commission of crimes. Confucianism, for example, or Buddhism does not appear to incite their adherents to crime easily or often. National Socialism, by contrast, and likewise Marxism, do easily and naturally hold out such incitements to their adherents, and indeed (as is obvious) owe a good deal of their attractiveness to this very fact.

It is impossible to deny that, in this respect, Darwinism has a closer affinity with National Socialism or Marxism than with Confucianism or Buddhism. Darwin told the world that a "struggle for life," a "struggle for existence," a "battle for life"[29] is always going on among the members of every species. Although this proposition was at the time novel and surprising, an immense number of people accepted it. Now, will any rational person believe that accepting this proposition would have *no* effect, or only randomly varying effects, on people's attitudes towards their own conspecifics? No. Will any rational person believe that accepting this novel proposition would tend to *improve* people's attitudes to their conspecifics—for example, would tend to make them less selfish, or less inclined to domineering behavior, than they had been before they accepted it? No.

Quite the contrary, it is perfectly obvious that accepting Darwin's theory of a universal struggle for life must tend to *strengthen* whatever tendencies people had beforehand to selfishness and domineering behavior towards their fellow humans. Hence it must tend to make them worse than they were before, and more likely to commit crimes: especially crimes of rapacity, or of cruelty, or of dominance for the sake of dominance.

These considerations are exceedingly obvious. There was therefore never any excuse for the indignation and surprise which Darwinians and neo-Darwinians have nearly always expressed, whenever their theory is accused of being a morally subversive one. For the same reason there is, and always was, every justification for the people, beginning with Darwin's contemporaries, who *made* that accusation against the theory. Darwin had done his best (as I said in Essay 2 above) to separate the theory of evolution from the matrix of murderous ideas in which previously it had always been set. But in fact, since the theory says what it does, there is a limit, and a limit easily reached, to how much *can* be done in the way of such a separation. The Darwinian theory of evolution *is* an incitement to crime: that is simply a fact.

It is perfectly possible, of course, and indeed it constantly happens, that publishing a certain proposition is an incitement to crime, and yet that the proposition is a true one. If a large amount of money, or drugs, or firearms, is unprotected at a certain place, and I publish this truth, then I incite to crime: indeed, "the greater the truth, the greater the incitement." This is merely an instance of what every sensible person knows: that there are truths which morally ought not to be told to children, to the moribund, to people whose sanity hangs by a thread, or to the criminally inclined. So I do not mention Darwinism's being an incitement to crime as a reason for thinking that it is false. I mention it as a fact worth knowing, which is almost never stated, but is, very often indeed, willfully concealed even by people who know it perfectly well.

BUT I *have* in this essay given a reason (additional to those I gave in Essays 2 and 4 above) for believing that the Darwinian theory of evolution if false. Namely, that it implies a struggle for life among humans which is far more severe, and a child mortality which is far higher, than any which really exists, or indeed could exist, consistently with our species surviving at all. At least in the case of our species, the unobservability of the struggle for life, which Darwin and Wallace made such ludicrous attempts to explain, has a very simple explanation: the struggle does not exist. *That* is why it has never been observed, either in civilized or in savage life.

There are countless populations of animals, as I pointed out in Essay 2, in which a struggle for life is quite certainly not going on. For example, the dogs which are domestic pets in a Sydney suburb, the racehorses in a large training establishment, or the merino sheep in a prized breeding flock. Not even the most armor-plated Darwinian will believe that *these* populations are constantly increasing up to the limit that their food supply permits, and are in consequence subject to severe child mortalities and to natural selection. The prize merinos, after all, are undergoing intensive *artificial* selection by their human controllers, and no sane person thinks that they are undergoing natural selection *as well*.

So when I say that in human populations the Darwinian struggle for life does not exist, it is not as though I am claiming for our species some privileged status which is unheard-of anywhere else in the animal world. I am merely claiming for us a status which exists, and which everyone knows exists, in countless populations of domestic or semi-domestic animals.

That there is a marked analogy, at the very least, between civilized humans and the domestic breeds of other animals did not escape Darwin's notice. He wrote that "civilized men . . . are in one sense highly domesticated."[30] The remark was somewhat cryptic, since Darwin does not say *what* sense this one is; nor did he develop the thought anywhere else, that I know of. But a great biologist of our own century, Konrad Lorenz, has fleshed out Darwin's hint with massive and compelling detail, in some pages which no one who has ever read them is likely to forget.[31]

Invaluable though those pages of Lorenz are, they also invite a very obvious and serious objection. He, like Darwin, is plainly ascribing a "domesticated" character, not to the members of our species in general, but just to the civilized ones among them. But if the word "domesticated" means what it says, then *by whom or what* can certain human beings have first been domesticated? By savage humans? By some other animals? By gods, or (as was once very widely believed) by demigods, such as Prometheus and Heracles? Every possible answer seems to be equally unsatisfactory.

Besides, this idea of Lorenz and Darwin puts too wide an interval between civilized and savage man. When Darwin first encountered the Yahgan Indians in their homeland of Tierra del Fuego, he was thunderstruck. "I could not have believed how wide was the difference between savage and civilized man: it is greater than between a wild and [a] domesticated animal, inasmuch as in man there is a greater power of improvement."[32] But in fact Darwin was mistaken about the Yahgans: indeed, just about as completely mistaken as it would have been possible to be. We know this from the testimony of a man who was born and spent most of his life among them. This was Lucas Bridges, whose parents were Christian missionaries to the Yahgans, and the first white settlers in Tierra del Fuego, only a few decades after the *Beagle's* visit.[33]

Darwin, and everyone else on the *Beagle*, believed that the Yahgans were cannibals. In fact they were so far from being so that, among them, even to eat the flesh of a condor earned opprobrium, because that bird *might* have eaten human flesh. Their language, Darwin wrote, "scarcely deserves to be called articulate."[34] But Lucas Bridges' father compiled a Yahgan dictionary of 32,000 words, which did not pretend to be complete at that. Darwin thought other Yahgans were careless and even brutal towards their children; in fact they were intensely devoted to them. Their religion was as important to them as it is to every primitive people; Darwin was unaware of its very existence. And as for their everyday social life, much of it was, even by Darwin's own description, "just like home." There was, for example, lying *and* detestation of liars; theft *and* recognition that theft is a crime; men who alienated the affections of other men's wives, amid general disapproval; and so on. All of which was no doubt equally true of the civilized people on board the *Beagle*, even under the command of that most formidable autocrat, Capt. Robert Fitzroy, R.N.[35]

Yet Darwin actually said, as we have just seen, that the difference between the Yahgans (say) and the people on the *Beagle* (say) is *greater than* the difference between a wild and a domesticated animal! Just a few weeks earlier, he had had a blazing row with Fitzroy, as a result of which he thought he might be obliged to leave the ship. Now suppose Fitzroy had marooned him among the Yahgans, with only the clothes he stood up in. There would then have been only two ways he could have survived: by being sustained, like a Yahgan child, by the generosity of the adults; or by getting food every day, as the adults did, without even the assistance of fish hooks or bows and arrows, in one of the most appalling climates and inhospitable terrains on earth. *Then* he would have seen the world rightly! In particular he would have seen rightly the difference between civilized and savage man. He would have seen that if the Yahgans were wild animals, or something more foreign still, then so was he; and that, since he clearly was not that, then neither were they. In other words, the domesticated character of human beings— although "domesticated" is plainly the wrong word—belongs to our species in general, not just to its civilized members.

As for that "struggle for life" among conspecifics, supposedly universal an constant, which Darwin was later to make famous, he saw nothing of it among the Yahgans. Well, that should go without saying: there was none of it there. Collecting shellfish, their commonest form of food-getting, was done by family groups, or by individuals. In winter, when the guanaco, with a good layer of fat on them, were forced downhill by the deepening snow above, a group of men would go off for a few days to hunt them. Whatever they got was simply shared among the hunters, who carried home as much as they could to share with their families. Well, they would, wouldn't they? Only someone who had "the struggle for life" on the brain would expect anything different.

Come to that, what *would* it be like, to meet a population of humans who really were always engaged in a Darwinian struggle for life? I cannot say. The best I can summon up is a very indistinct picture of a number of people in a sort of pandemonium competition for food. In my picture, the people are not distinguishable from one another by age, by sex, by rank, or by anything—although even gulls, even when fighting over food, are still differentiated to some extent by their age, sex and rank! As the word "pandemonium" will suggest, the figures themselves, in my picture, tend to diverge from the human form, in the direction of cruel beaks, or claws, or teeth. I would not bother the reader with these introspection-reports, if I believed for one moment that anyone else has, or ever has had, a more plausible idea, of what a Darwinian struggle for life among humans would be like. But I do not believe that anyone has.

What would it be like, even, to meet a man who really *believed* that there is a Darwinian struggle for life among humans? Even this, as far as I know, never happens, and never has happened. Which is certainly a very great mercy. But it is not at all difficult, on the other hand, to *imagine* meeting such a person.

He would be a man who actually believed that he is *struggling for his life*, all the time, against his parents, children, wife, neighbors, the postman, the doctor, the Lord Mayor . . . , and also believes that everyone else is in exactly the same case. What could we possibly make of this most unfortunate man? His mental state, because of its obvious affinity with certain more familiar al states, might aptly be named *paranoia darwiniensis*. In any case it would be clear that he is in some extremely dreadful delusionary state. Nor would any cure seem at all possible, unless it began with someone's convincing him that Darwin's theory of evolution is false.

Tax and the Selfish Girl, or
Does "Altruism" Need Inverted Commas?

. . . it is, after all, to [a woman's] *advantage* that her child should be adopted.

—R. Dawkins, *The Selfish Gene*

I

THERE ARE SOME BELIEFS which, though we have found them to be false over and over again, never entirely lose their hold on us, because they appeal to something permanent in human nature. An example is the belief that "the grass is greener on the other side." Everyone of mature years knows this to be false, and yet it always retains some degree of influence on our behavior. Indeed, this particular delusion is so deep rooted that it is not even confined to human beings. I have actually seen a cow escape from the well-grazed paddock in which she had long been kept, and promptly put her head back through the wire fence and begin grazing *inside* her former prison.

There are other beliefs which, though disproved countless times, never die out, because they appeal, not to something in everyone, but to a certain perennial type of person. An example is the belief that everyone is at bottom selfish, or that no one ever acts intentionally except from motives of self-interest.

There *is* a perennial human type to whom this belief is peculiarly and irresistibly congenial. It is almost never a woman. It is a kind of man who is deficient in generous or even disinterested impulses himself, and knows it, but keeps up his self-esteem by thinking that everyone else is really in the same case. He prides himself both on having the perspicacity to realize, what most people disguise even from themselves, that everyone is selfish, and on having the uncommon candor not to conceal this unpleasant home truth. Who has not met people of this type? In the Australian-English of fifty years ago, there was a wonderful expression for this kind of man: he was said to be "as flash as a rat with a gold tooth."

This "selfish theory of human nature" has a long history, but not, on the whole, an impressive or even a respectable one. It has nearly always led a kind of underground life, finding favor much more often with those who are ignorant and envious, than with either the educated or the privileged. It comes into its own, however, in periods of Enlightenment, such as the fifth century B.C. in Greece or the eighteenth century A.D. in western Europe; and the reason is obvious enough. Once the belief in gods has evaporated, it becomes at once a most pressing question how to account for the prominent part played in all previous human history by the supposed human representatives of gods: priests and kings.[1] And the explanation which comes most naturally to an Enlightened mind is that priests and kings had been, initially at least, simply impostors: ordinary people to whom some opportunity had been given, by the ignorance or weakness of their fellows, to profit at the expense of others, and who, being selfish, had naturally taken full advantage of this opportunity.

The belief in evolution (as I said in Essay 2) was itself peculiarly a product of the eighteenth-century Enlightenment, and from that circumstance alone would have had some affinity with the selfish theory of human nature. But the *Darwinian* theory of evolution has an especially strong affinity of its own with the belief in universal selfishness. For Darwinism says, after all, that in every species the individual organisms are always engaged in a struggle for life with one another. And what could that struggle be, except a school in which the scholars do well in proportion as they are ruthlessly selfish? An organism or a lineage of organisms which was unselfish or altruistic—which ate less or mated later (for example), in order that some conspecifics could eat more or mate sooner—would hardly be going the right way about becoming a "*favored* race in the struggle for life." It would, on the contrary, be taking the shortest possible path to being eliminated by natural selection, which can never sleep, and can never forgive inferior performers in the game of survival and reproduction.

Logically, therefore, Darwinians ought always to have accepted the selfish theory of human (and all animal) nature: accepted it from the very start, and openly. But nothing at all like that happened in fact. No doubt part of the reason was that Darwin (as we have seen) had quite enough to do in 1859 as it was, in the way of prizing evolutionism loose from its accustomed setting of republicanism and anti-religious zeal. If he had openly embraced the selfish theory in *The Origin of Species*, it would have sunk the book like a stone; and Darwin would have been very well aware of this fact.

But he was probably never even tempted to do so. The selfish theory requires, in those who believe it, an appetite for insolence and absurdity far stronger than most people possess or even approach. It is a doctrine tenable only by Hard Men, or (as the French say) *esprits forts*. Insolence held no charms whatever for his emollient temperament, and anyway he had too much common sense to believe the selfish theory.

In this respect Darwinians, both in the nineteenth and in the twentieth centuries, have nearly always stayed within the lines laid down by Darwin's personal common sense and caution. They have admitted, very often indeed, that altruism, and especially altruism on the scale that human beings go in for it, is a "Problem" for their theory of evolution: that is, is inconsistent, *prima facie* at least, with that theory. Which is fair enough, as far as it goes. But some of us well remember Hume's reasonable protest against the "custom of calling a *difficulty* what pretends to be a *demonstration*, and endeavouring by that means to elude its force and evidence."[2] And in fact Darwinians have, beyond all doubt, cruelly overworked this trick, of calling altruism a "problem" for their theory, and then proceeding to think and write as though there were really no such problem at all.

Darwin himself never even conceded that altruism *is* a problem for this theory of evolution. He was not in the habit of drawing attention to problems for that theory, unless they were ones which he believed he could solve. He appears never to have been worried by the obvious disadvantages under which altruistic individuals would lie in "the struggle for life." Instead of that, and characteristically, he explains to his readers at considerable length that altruism would actually be an *advantage* to a tribe of humans, if it had to compete with another tribe which, though otherwise equal, was less given to altruism among its members.[3] This is true, obviously enough; indeed, a good deal more obviously than enough. But it does nothing at all to explain how, in a constant competition among conspecifics to survive and reproduce, altruistic individuals could possibly avoid being demographically "swamped" by non-altruistic ones.

Most Darwinians, then, since ever there have been Darwinians, have resisted the strong "gravitational pull" exerted on their minds by the theory that humans and all other animals are selfish. In the last twenty-five years, however, this situation has radically changed. The selfish theory has been openly, and in many cases aggressively, embraced by a large and influential group of neo-Darwinians: the ones who have come to be called "sociobiologists."

These people are nothing if not *esprits forts*. "Scratch an 'altruist' and watch a hypocrite bleed," wrote one of them.[4] "Nice Guys Finish Last" was the expressive title of a representative article by another.[5] A third writes that "altruism [is] something which has no place in nature," or in human nature either: "we are born selfish."[6] All transactions between organisms, no matter how altruistic some of them may appear, are in reality (according to these thinkers) cases of one organism manipulating another for its own advantage: even the transactions between parents and their children. Sociobiologists all agree that "[natural] selection would favor parents who succeeded in manipulating their offspring, over those who did not."[7] In fact these authors think, and say, that you will be on the right track in biology, if you expect to find "dirty tricks"[8] and "dog eat dog"[9] *everywhere.*

Sociobiology is like garlic: a little goes a long way. So the above quotations, few and short though they are, will probably be enough to convey the essential flavor of sociobiology to any reader previously ignorant of it. Even in these glimpses, you can see the tell-tale gold tooth sparkling a mile off. But it will be helpful to mention some less summary expressions of the selfish theory as it has been revived by these thinkers.

There are physiological or behavioral signals of submission which in our species, in dogs, and in many other animals, terminate fights between conspecifics, or prevent them from starting, or at the least usually prevent them from ending in a death. The existence of these signals, according to Professor E. O. Wilson, the leader of the sociobiological school, is profoundly puzzling. They constitute, he says, "a considerable theoretical difficulty: Why not always try to kill or maim the enemy outright?"[10] This scholarly enquiry might well cause you, if you are a mere normal man, and can remember being in a school playground fight or two, a sharp intake of breath. But if, of course, you are a Darwinian, and believe that all organisms, including yourself, are engaged in a struggle for life, or if you take for granted that humans and all other animals are selfish—why *not*, indeed, "always try to kill or maim the enemy outright"?

As a second example, consider communication. Everyone knows that organisms sometimes communicate with one another as part of an attempt at manipulation of the "signal-receiver" by the "signal-sender." An unscrupulous second-hand car salesman, talking to a potential buyer, is a stock example of such self-interested communication. So is Brer Rabbit, when he pleads with

Brer Fox *not* to throw him into the briar patch (where in fact he lives and thrives). But according to the sociobiologists R. Dawkins and J. Krebs, *all communication whatever* is "manipulation of signal-receiver by signal-sender."[11]

As a third example, consider the phenomenon of "baby snatching." Among certain species of monkeys, as well as among ourselves, it sometimes happens that a bereaved mother will steal another mother's baby, "adopt" it, and care for it. Most people have heard of this phenomenon, and nearly all those who have—certainly not just the bereaved mothers themselves—feel in a dim way, and with a dull pain, that they *understand* it too. Not so the sociobiologist Dr. R. Dawkins. He finds the fact of baby snatching deeply and importantly puzzling; as well he might, given his Darwinian assumptions. As Dawkins sees the matter, "the adopter not only wastes her own time: she also releases a rival female from the burden of child-rearing, and frees her to have another child more quickly. It seems to me a critical example which deserves some thorough research. We need to know how often it happens; what the average relatedness between adopter and child is likely to be; and what the attitude of the real mother of the child is—it is, after all, to her advantage that her child *should* be adopted; do mothers deliberately try to deceive naïve young females into adopting their children?"[12] If they don't, *why* don't they? A question, I need hardly say, even more breath-taking than Wilson's question about ignoring submission signals in a fight.

I hasten to add, in order to be fair to Wilson and Dawkins, that they, in marked contrast to some other sociobiologists, actually *approve of* human altruism. Far from writing about it with cynicism or even incredulity, they make it quite clear that they think there should be more of it.[13] Well, according to their own account, there could not possibly be less, since there could not be any at all. We can therefore only ascribe these authors' enthusiasm for altruism to an amiable inconsistency on their part. Either to that, or to their attempting to manipulate their readers for their own advantage. For Dawkins is one of the sociobiologists whom I mentioned a moment ago, as believing that *all* communication is self-interested manipulation. And then, a fellow sociobiologist whom Dawkins admires[14] has been candid enough to say in print that "morality aside, the optimum strategy for the unabashed egotist is *unwavering praise of altruism*."[15] So perhaps Dawkins' praise of altruism is not an amiable inconsistency after all, but something more consistent, and less amiable.

Even on that supposition, however, it is a mystery what the writer just quoted can have meant by his proviso, *"morality aside."* All moral education is simply some more self-interested manipulation, if the sociobiologists are right. So what possible need can there ever be, or indeed what would it even mean, to set it "aside"? But this difficult question is plainly one which is best left to specialists in the exegesis of the New Darwinian Testament.

II

IF PROFESSOR WILSON were right, it would be a "considerable theoretical difficulty" why Darwin did not try to kill or maim Samuel Butler, for example, or why Wilson himself does not try to kill or maim his bitter enemy and Harvard colleague, Professor R. C. Lewontin. But this is *not* a considerable theoretical difficulty. It is just a joke, and a stupid one at that. It would probably be recognized as such even at Harvard; though it would certainly be recognized there as a risky joke too.

Dr. Dawkins, likewise, cannot understand why mothers do not welcome baby snatchers, and says that the question "deserves thorough research." (This phrase is, of course, academese-English for "I have got all these unemployed graduate students . . . .") Though this is a perplexity which few can share, Dawkins is not absolutely the first to perceive the difficulty. In *Uncle Tom's Cabin* the slave Eliza flees with her young son, in order to prevent his being sold off the plantation, and the slave dealer is thereby put to the expense of hiring slave hunters to capture the runaway pair. The dealer and the hunters ruminate on Eliza's perplexing behavior as follows. "This yer young-un business makes lots of trouble in the trade . . . If we could get a breed of gals that didn't care, now, for their young uns . . . "'twould be the greatest mod'rn improvement I knows on . . . I never couldn't see into it. Young uns is heaps of trouble to 'em—one would think, now, they'd be glad to get clear on 'em; but they aren't."[16]

Since Dawkins likewise cannot "see into" mothers who do not welcome baby snatchers, the sensible place to begin the research he wants done would be by asking his own mother why she did not offload *him*? Unless she too has been unhinged by the Darwinian vision of human life, as a ruthless competition to survive and reproduce, her answer would be something like the following. "Useless as it may now appear from your present 'scientific' perspective, and perhaps from certain other perspectives too, *I preferred not to.*" But even an imaginary scene is painful to contemplate when it includes so much of both absurdity and insolence; let us draw a veil . . .

"WHY DON'T MEN or dogs in a fight always try to kill or maim their enemy?" "Why don't human or monkey mothers welcome baby snatchers?" These questions are typical expressions of the selfish theory of human (and all animal) nature, as it has been revived by neo-Darwinians in the last twenty-five years. Before the late 1960s, no one had ever asked, or thought of asking, questions like these.

But they are also, once you stop to look at them, questions typical of the new era in Western civilization which began in the late 1960s. They have an unmistakable family resemblance to many popular questions which bear the peculiar impress of the last twenty-five years.

"Why didn't all young American men eligible for the draft, instead of just thousands of them, flee their country to avoid fighting in the Vietnam war?" "What if they gave a war and nobody came?" "Why shouldn't a woman have as many abortions as she wants to?" "What right has the government to steal part of my money, and call this theft taxation?" "How *can* the Pope continue to oppose contraception? Can't he see that overpopulation is ruining the environment everywhere?" It would be easy to multiply examples of these characteristic questions of our age; but their family resemblance is so obvious, that it can hardly be necessary.

These typical questions of our age are all foolish, and foolish in the same way as the typical questions of sociobiologists. Why does not a monkey or a human mother offload her babies for her own advantage, indeed! A feminist might just as sensibly ask a termite queen, why she does not in her own interests break out of her prison, do something about her terrible figure, and start reading the most emancipated female authors. A draft dodger might just as sensibly ask an American soldier ant why he, too, does not run away to Canada when war threatens his survival.

The folly which is common to the favorite questions of our time, and to the typical questions of sociobiologists, lies in a certain presupposition which they have in common. This is, that human life, and indeed all animal life, is best understood by comparing it with the model furnished by youngish American adults of the last twenty-five years. By people, that is, who are, beyond all historical precedent, free, rich, mobile, innocent of the very idea (let along the reality) of food shortage, under no necessity to work, unburdened by familial, religious, or other loyalties, undistracted by education, curiosity, or any disinterested passion, principally anxious (if male) to preserve a whole skin, and (if female) to preserve her immaternity. They (as the saying is) "just want to have fun," and are the first instance in history of an entire generation, as distinct from a tiny minority, being in a position to realize this challenging idea. For this very reason, however, they make a highly misleading model for human biology; and a still worse one, for general biology.

It is because of this shared though tacit ideal of human and animal life, that the draft dodger and the sociobiologist both wonder how soldiers can be found for human wars, or for ant wars; that the sociobiologist and the ruthless free marketer are both puzzled by the effectiveness of submission signals in human fights and dog fights; that the feminist and the sociobiologist are both mystified by baby snatching among women and monkeys;

that the anti-religious zealot and the sociobiologist both marvel at the fact that a celibate priesthood can be kept up; and so on.

All of these shared puzzlements are special cases of one general puzzlement, common to the sociobiologists and to all our representative contemporary questioners: *why don't organisms behave with more regard to their own advantage—that is, more selfishly—than they do?* Of course no two of these various groups have exactly the same conception of "advantage." The sociobiologist identifies advantage with increased chances of survival and reproduction; the draft dodger identifies it with increased chances of survival, and cares nothing at all about reproduction; the feminist identifies it with increased chances of survival without reproduction; the free marketer identifies it with increased chances of survival and enrichment. But none of the various facts which puzzle these people would puzzle them at all, if they had not assumed from the outset that humans and other animals *are* all selfish.

And yet it is plausible enough, or at least as plausible as Darwinism in general is, that a monkey mother *would* have more descendants if she offloaded each baby in turn onto someone else; that dogs who ignored submission signals from their opponents would have more descendants; that soldier ants who escaped the next would (if they remembered to carry off a female or two) be a bigger biological success than the G. I. Joes who stayed at home. And yet, well, they just *don't* do these things. You can easily see what the source of the trouble is. These wretched dogs, monkeys, ants, etc., simply do not know their own neo-Darwinian business; although one would have thought that the imperative "Survive and Reproduce!" was unambiguous enough. At any rate, they do not know their own business as well as sociobiologists do.

III

HUMAN SELFISHNESS goes very deep and extends very far. But that is obvious, and not in dispute. It needs no expensive education in biological science to teach us *that*; nor did we have to wait to learn it from the recent examples of draft dodgers, feminists, or the business virtuosos in dog eat dog and dirty tricks. The question is, whether there is not also an opposite side to human beings—an unselfish or altruistic side—which also goes very deep and extends very far. The sociobiologists say there is not. I say there is.

Human beings do not like to be absolutely alone for very long, or to have the company only of members of other species. Almost all the higher mammals feel the same way. If you put a number of members of any one of these species into a certain large area, they will almost never distribute themselves over the area in a random way, but will always "cluster" to some extent. If two of these species are put into the same large area, it will almost always be found that every animal is closer to some conspecific than to any member of the "foreign" species.

It is easy to think of a Darwinian explanation for this common mammalian preference for the proximity of a conspecific: so easy as to be uninteresting. (Given a certain biological fact to start with, it usually *is* so easy as to be uninteresting, to think of a Darwinian explanation of it: that is one of the many sources of rational dissatisfaction with Darwinism.) Here, however, it is the fact which is of importance, not its explanation, whatever that may be. A preference for the proximity of at least one conspecific is not altruism, of course. Indeed, it is not even sociability: only the first approach to it. But very obviously, if it did not exist, there would be much less scope for altruism than there actually is.

Anyway, the proximity of a conspecific is practically necessary in order for us to indulge the strongest passion of our nature: for *communication*. There are few punishments which we dread as much as solitary confinement. And yet, how amazingly difficult it has always proved to be, for prison authorities to prevent all communication even between the inmates of solitary confinement cells! People *will* talk. They will talk to themselves, to the cat, to their hat, to a post, if there is no other person to talk to. They will talk with their fingers if they are deaf and dumb, and with gestures among foreigners, if that is the best they can do. They will talk with their eyes, at the last, when they can do no more.

More specifically, human beings love to *teach*. It is the young, of course, who are the main recipients of teaching, but in fact nearly all of us are strongly disposed to teach nearly anyone whom we take to be our inferior in knowledge, skill, or *savoir faire*. This impulse shows its strength very early, especially in females. It is scarcely possible to prevent a four-year-old girl from "teaching" her young siblings or her dolls what she herself has lately been taught. In most people, the propensity to teach becomes even stronger with time: that is one reason for the garrulity of age. Correspondingly, there is an innate and strong propensity in the human young to *be* taught. Everyone would recognize that a boy of five (say) is in some profoundly pathological condition, if he shows no tendency at all to learn anything from other people. And there is nothing in any other species (it can hardly be necessary to say) remotely approaching the human passion for teaching the young. In every other species, protection from danger, and provision of food, exhaust or near enough exhaust what adults do, or can do, to help beginners.

The intense communicativeness of our species must have evolved; but this pious formula only amounts to the triviality, that we have it now and that not all of our ancestors did. If Darwinism is true, our communicativeness must have come down to us because it was advantageous to people, or to some of the ancestors of people. But once again, it is the fact of our communicativeness which matters here, not how it is that we come to have it.

Anyway, it is perfectly obvious that people do not *now* communicate, or communicate as much as they do, because of any advantage which accrues to themselves from communicating. Indeed, there are few human experiences more common than that of people finding that they have *injured* their own interests, by too great a readiness to communicate, or too great a receptivity to the communications of others. Yet lessons of this kind are constantly thrown away on us, simply because our love of communication is so strong, and so little controlled by a regard for our own interests.

While there is nothing in the least self-interested about the great bulk of the communicating we go in for, there is nothing altruistic about it either. We talk and write, show photographs, lend books or music, send anonymous threats, etc., neither for our own advantage in general, nor for anyone else's. Advantage does not come into it. With us, communication is an end in itself.

Of course our communicativeness is not itself altruism, any more than our preference for the proximity of a conspecific is. But, again like that preference, it is a practically necessary condition for the exercise of altruism. And not only that: the connection between our communicativeness and our altruism is in fact extremely close. For it is perfectly obvious that there would be vastly less communication among humans than there actually is, if the sufferings of other people were a matter of indifference to us (except as a danger signal for our own case) or were a source of positive satisfaction to us. Communication is *repressed by* indifference towards the sender of a message. So it is by enmity towards the sender: that is why messengers have a better than average chance of being shot.

For millions of our contemporary human beings, the enormous size of our cities has merely served to provide first-hand experience of the unparalleled misery of loneliness. To these unfortunates, the minimal or entirely imaginary communication afforded by a cat, or even a goldfish—nay, by a pot plant—is, often enough, some slight balm. What a poignant commentary on the depth of our hunger!

Once a human society reaches a certain size and complexity, it always develops (as I said in Essay 2) groups which discharge a specialized social function: most notably, the military function, the religious, and the medical. Now, the existence of such special groups of people as soldiers, priests, and doctors is inconsistent, at least on the face of things, with the selfish theory of human nature. A soldier, after all, takes on himself, in order to defend his society against hostile societies, a higher risk of wounds and death than non-soldiers are exposed to. A priest takes on himself, in order to protect the society from divine displeasure, permanent or periodic sexual abstinence, fasting, self-mortification, etc. A doctor takes on himself the task of trying to improve the condition of the sick, injured, or dying: an occupation which but few humans find positively attractive, and which one would expect, if the selfish theory were true, to be invincibly uninteresting to everybody. Nor do soldiers, priests, or doctors in general receive any biological *advantages* over others, which might compensate for these biological penalties which they take on themselves.

The non-specialized members of a human society are nearly always conscious of the debts that the society owes to its specialist groups. Everyone recognizes, more or less distinctly, and feels some gratitude for, the unusual subordination of self-interest of others which is required by the profession of arms, of religion, or of medicine. These professions, unlike those of (say) cooks or prostitutes, are *respected*. The specialized groups themselves, by an internal dynamism of their own, always proceed to make their own altruistic contributions to society more conspicuous still. The army develops an élite corps, distinguished from the rest by exceptional bravery and discipline in war. The priesthood throws up a certain suborder of priests, distinguished by a higher degree of sanctity and self-denial. The most valued doctors, tending to be monopolized by the most important patients, insensibly develop into a "royal college of medicine," which is authorized to dictate standards to the generality of the medical profession.

People, I have said, "take on themselves" the biological penalties inseparable from being a soldier, priest, or doctor. The phrase is a little too suggestion of contemporary "career choice," but it is nevertheless essentially right. Membership of the army, priesthood, or medical profession has occasionally been hereditary, but usually it has not. It has never been determined by a random process, or by a democratic process, or by some somnambulistic one. Joining one of these specialized groups has nearly always been not only a voluntary step, but an informed one. That is, people *know in advance* that, as soldiers, they increase their chances of suffering wounds and death; that, as priests, they are expected to forfeit some or all sexual enjoyments, certain foods, hours of rest, and so on; that, as doctors, they will often be required to subordinate not only their own convenience, but the care of their own health to caring for the health of other people. No doubt they do not usually know in advance, as well as they will know later on, what they have "let themselves in for." But then, *that* is true even in the most sophisticated and information-saturated societies.

The professions of arms, of religion, and of medicine, are therefore essentially inimical to the self-interest of the individuals who compose them. And yet—to borrow Wittgenstein's famous phrase—*these games are played*. They are never played nearly so well as the official ideals of the respective games prescribe; but that is trivial, because no game ever is. There perhaps never was in the world a soldier, priest, or doctor who did not sometimes passionately wish that his profession was anything other than what it is. Well, these *are* specialized jobs: not everyone is suited to them, and sometimes it seems as though no one is suited to them. Sometimes it is hard to recruit potential priests; at other times no one wants to be a doctor; and armies sometimes prove impossible to fill on the voluntary principle. Yet somehow or other, under every vicissitude of history and every fluctuation of culture, these three specialized professions *are* kept up, and their essentially altruistic games are played. And what, after all, could be more natural? Or rather, what *is* more natural in human life, than the existence of an army, a priesthood, and a medical profession?

I hope no reader will suppose I need to be reminded that sometimes soldiers run away before a battle begins, that some priests are selfish hypocrites, or that some doctors cynically extort money from helpless or terrified patients. All of that Enlightenment stuff goes without saying, but it is quite irrelevant here. My purpose is simply to argue against the selfish theory of human nature, and it would make no difference to my argument if in fact soldiers ran away nine times out of ten. But for selfish theorists, and hence for sociobiologists in particular, it is the fact that soldiers *do not always* run away, and that priests and doctors *sometimes* approximate to the official ideals of their profession, which constitutes "the problem of altruism," or at least a part of it.

Of course if soldiers always did run away, and if priests and doctors were always selfish impostors, then sociobiologists would be able to explain these facts with supreme *éclat*. All these organisms would simply be doing what Darwinian theory predicts: increasing their own chances of survival and reproduction. But alas, as things actually stand, human soldiers, priests, and doctors—like ant soldiers, monkey mothers who prefer to keep their babies, and dogs that accept the submission signals of opponents—seem determined to deny sociobiology all the most resounding explanatory triumphs that it might have enjoyed.

IT IS WORTHWHILE to take a closer look at one of the three professions I have been speaking of. Doctors are seldom inclined to under-value the contribution which they make in their professional capacity to the well being of others. The same is true of almost every branch of the medical profession, and the ward sisters in hospitals are certainly not an exception to this rule. They are much more likely to over-estimate than to under-estimate the extent to which the welfare of patients depends on their discharging their professional duties conscientiously.

In every large hospital, however, there is always one ward sister who, for her professional self-esteem, puts all the other sisters in the shade. It is always the same one: the maternity ward sister.

This fact is not generally known. But among experienced hospital workers, of whatever department, it is a fact so familiar as to go entirely without saying; like (say) the diffidence of doctors just out of medical school. Any other sister in the hospital meets from time to time with some check to her authority to which she is obliged to submit. But the maternity sister exercises, in her domain, an authority which virtually no man born of woman—"no, nor woman either"—ever challenged with success. If you tangle with her, then you had better (as the saying is) hang on to your night job.

The unique professional self-esteem of maternity ward sisters is not only known to experienced hospital workers as a fact: the reason for it is just as well known as the fact. The reason is this: that the maternity sister, and she alone in the hospital, *saves lives every day*. Everyone else, from the medical superintendent down to the humblest clerk in "Stores," can sometimes fairly claim to have made some contribution to the hospital's altruistic goal of saving lives. But the maternity award sister is something entire apart. *Her* altruistic bent is like Falstaff's dishonesty—"gross as a mountain, open, palpable," constant, incorrigible. It is what she is *paid* for, as plainly as boxing is what a professional boxer is paid for; and like him, though unlike many others, she *does* what she is paid to do.

But the most instructive thing about the maternity sister is neither the fact of her unique professional self-esteem, nor the explanation of this fact. It is this: that *no human being really disagrees with* her own estimate of her value. You may think that you disagree with it. You may be one of those people who have the "population explosion" on the brain, and declaim with passion, in public and at home, against the appalling curse of human fecundity. But if you were admitted to the delivery room, and watched the maternity sister at work, you would soon find that a new set of priorities was emerging among your values. This would happen in any case, but especially so if the delivery threatened to be a dangerous one for the child or mother or both. Your overriding wish, you would then soon find, is the same wish as the sister brings to her work every day: that the mother and the child should both, so far as the situation permits, come well out of it.

What are we to make of the maternity sister, and of the respect in which she is held not only by herself but by all of us? Even for neo-Darwinian biology in general, she is at the very least a serious problem. For it is not, of course, as though she is *related to* her patients. Indeed, if it does turn out that one of her patients is a relative, she will not willingly perform the delivery. The ethics of her profession, which forbid her to deliver a relative except in an emergency, merely second the promptings of her own nature on this point.

But for sociobiologists in particular, and indeed for anyone who inclines to the selfish theory of human nature, the maternity sister is something a good deal worse than a "problem." If it is (as we have seen it is) a problem for sociobiologists, demanding "thorough research," why women do not offload their infants onto other women, what shall we say of the maternity sister? She *constantly and deliberately extends her best efforts to facilitate the reproduction of women to whom she bears not the slightest relation of heredity.* She is not even paid, at that, any fancy salary for doing so: far less than many other people in the hospital are paid. But whatever she was paid, why would she do it, when she might instead be reproducing *herself, and* using all the dirty tricks which her expertise might suggest, to *repress* the reproduction of "rival women"—just as sociobiology predicts?

A question for sociobiologists to ask themselves, indeed; and one which will require, no doubt, a vast amount of "thorough research." A better question, though, would be this one: "How many ridiculous pseudo-problems is your theory allowed to generate, before it is time to conclude that the only real problem in the case is *your theory?*"

IV

In the same spirit, it is time I stopped saying that the specialized groups of people I have been talking about are a "problem" for the selfish theory of human nature, or that they are *"prima facie* inconsistent" with it. *Are* they inconsistent with it, or are they not?

Suppose they are. Then the selfish theory is so obviously false as to be not worth talking about. The motivation and behavior of maternity sisters, of soldiers who are loyal and brave (and so on), are not out of the way facts, or obscure facts, after all. They are specialized facts, indeed, and different from the motivation and behavior of most humans most of the time. But they are perfectly obvious facts, and therefore any proposition which is inconsistent with them is obviously false. On the present supposition, then, the theory that people are really—when you come right down to it—selfish, is about as credible as the theory that people really, when you come right down to it, have two heads.

So let us try the other supposition: that the selfish theory is *not* inconsistent with the obvious facts which I have adduced about soldiers, priests, and the rest. In this case, the selfish theory does not deny the altruism of (say) maternity sisters or heroic soldiers.

But then, what *does* the theory deny, that anyone else affirms? It could only be some super-*degree* of altruism, or some super-*extent* of altruism, which no one has ever contended for the existence of. If the motivation and behavior of maternity sisters and heroic soldiers are "selfish" enough to satisfy the requirements of the selfish theory, then those requirements must be utterly trivial. They would be satisfied, in that case, even if *everyone* were as altruistic as heroic soldiers and maternity sisters. But at this rate, it will be obvious, the selfish theory has no content at all, or virtually none.

The selfish theory, then, either asserts something which is obviously false, or it asserts only a trivial truth. And as a matter of historical fact, the theory *has* always displayed a characteristic oscillation between these two "faces." It does so not only at the hands of its ignorant expositors, but at the hands of its learned ones. At one moment, the selfish theory appears to be a shocking denial of common knowledge and cherished values. Yet at others, it seems to shrivel into a truism, or even a tautology, which no one ever dreamt of doubting or denying.

You can find this oscillation going on, if you have antiquarian interests, in such famous authors of the selfish persuasion as Hobbes and Mandeville. If you prefer a contemporary instance of it, you can find it in (for example) the sociobiologist G. Hardin.

He lately defended the selfish theory as being a profound biological truth. But he could not resist also plaintively suggesting that it is really a most inoffensive platitude, amounting only to "a truism," "that winners win,"[17] and asserting only *what could not be otherwise.*[18] How one and the same proposition could *be* both a profound biological truth and a truism like "winners win," was not explained by Hardin; and would, indeed, require a mind more powerful than his to explain.

Critics of the selfish theory long ago traced its characteristic oscillation, between shocking falsity and trivial truth, to a specific pair of propositions. One is, that no one acts intentionally except from motives of self-interest. This is the proposition which gives the selfish theory its required element of insolence and arresting falsity. The other proposition is, that no one acts intentionally except from some motive or interest of *his*. This gives the selfish theorist the element of impregnable truth which he also requires. And this proposition is plainly, indeed, impregnably true; but only in the same way as any other tautology is. *Of course* no one can act intentionally except from some interest which he or she has; but this proposition is just as trivially true as (say) "No one can be married to anyone except his or her spouse."

It plainly does not follow from the tautology, "No man can act intentionally except from some interest which he has," that "No man can act intentionally except from motives of self-interest." That inference belongs to exactly the same class as the atrocious (though ever popular) inference from "Whatever will be, will be," to "All human effort is ineffectual." And to the same class (to borrow an example from Hume) as the inexcusable inference from "Every husband has a wife" to "Every man marries." If you set out from a tautological premise, you cannot validly infer from it *any* conclusion which is not itself tautological.

The possible objects of a man's interests, motives, or desires, are inexhaustibly various. Among them may be, to wreak revenge on a particular man, to gain the love of a particular woman, to serve his country, to experience the love of God, to understand contemporary physics, to witness the sufferings of others, to relieve the sufferings of others, to acquire money, to write a great book . . . the list is endless. But it can perfectly well happen, and often does happen, that a man pursues one or more of these "particular passions" without regard to his own interests; indeed, to the manifest injury of those interests, and even to the destruction of his wealth, health, or life itself. The man who, in defense of his good name, challenges another to a duel and is killed, is an old stock example; but still a good example nonetheless.

In the last three paragraphs (as any philosopher will know), I have been paraphrasing part of a classic work: the examination of the selfish theory by Joseph Butler in 1726.[19] In the fifty years or so before that date, Hobbes, Locke, and Mandeville, among others, had made the selfish theory immensely popular. By the time that Butler wrote against it, it had become, as he complains, an opinion *de rigueur* for every citizen who wished to be considered worldly wise, and not the "gull" of priests or other self-interested manipulators of their fellows.

The selfish theory (as I implied earlier) flourishes always and only in periods of Enlightenment. The first victims of Enlightenment, and the most important ones, are (of course) priests. Behind "monkish impostures"—pretensions to miracle-working powers, divine authority, extraordinary sanctity, etc.—Enlightenment discloses, or claims to disclose, nothing but worldly and cynical clerics looking after their own interests. The next victims of Enlightenment, and the next most important ones, are kings: especially kings in their martial capacity. Behind national military glory, Enlightenment discloses, or claims to disclose, nothing but the cupidity and vanity of crowned bullies.

By about 1725, then, the selfish theory of human nature was almost as much a part of the "Revolution Settlement" of 1688 as a limited monarchy, a Protestant succession to the throne, and a "church by law established." But alas, Enlightenment is not only irreversible but insatiable, and proceeds inexorably to devour its own children. It cannot stop with the "unmasking" of priests and kings as being entirely selfish: it must proceed to disclosing that *everyone* is entirely selfish. That is precisely what Hobbes and

Mandeville, for example, claimed to disclose in *Leviathan* (1651) and *The Fable of the Bees* (1714) respectively. These books, naturally, met with widespread denunciation by representatives of religion or of established governments. But they also met with even more widespread private agreement.

Enlightened opinions are *always* superficial. If priests, kings, soldiers, doctors (and so on) were nothing more than Enlightenment can see in them—if they were, in plain English, confidence men—then virtually the whole of human history would be unintelligible, and the question "what if they gave a war/a religion/a monarchy, and nobody came?" would be a sensible one. But that is not at all the way things really are.

THE SELFISH THEORY, as well as having the superficiality which belongs to all Enlightened opinions, has in addition a kind of unreality which is all its own. This has often been noticed: Mary Midgley, for example, accurately describes it as a *melodramatic* quality.[20] The reason is, that a selfish theorist is always, and inevitably, "playing to a gallery"—the gallery of the *un*selfish—which, according to the selfish theory itself, must be empty.

When a man is detected in acting selfishly, where it had previously been thought that he was acting altruistically or at least disinterestedly, we think the worse of him. We disapprove both of his selfishness itself, and of the hypocrisy or self-deception which had previously concealed it from others or from himself. The same thing happens when a whole class of men, priests for example, is detected in selfishness disguised as something else. An adverse moral judgment is passed on all members of that class, by everyone who is not a member of it. And this judgment is passed, again, both on their selfishness and on the hypocrisy or self-deception by which they had prevented it from being revealed earlier.

It is a very familiar function of satirical literature to reveal disguised selfishness, either in a particular man or in a whole class, and to elicit moral disapproval of it. Molière's *Tartuffe* and Samuel Butler's *Hudibras* are classic examples. In cases like these, and in all common cases, there is no difficulty in knowing to whom the revelation of selfishness is being made, or whose moral disapproval of it is being elicited. It is all of us who are not hypocritical and worldly Roman Catholic priests, in the case of Tartuffe, or (in the case of *Hudibras*) all who are not self-serving Puritan zealots. These make up in each case the gallery that the satire is being "played to."

The same satirical intention is present, of course, and is very obvious, whenever a writer advances the selfish theory of human nature. It is intended to detect selfishness where it was undetected before, and to elicit moral disapproval, both of it and of the hypocrisy or self-deception which had prevented its being revealed earlier. In this case, however, there is a difficulty, and a glaring one, about what gallery it is that is being played to. *To whom* can the revelation of selfishness be being made, and *whose* moral disapproval of it is being solicited? If every human being *is* thoroughly selfish, to whom can this fact come as a revelation, and whose moral disapproval is the revelation of it supposed to elicit?

The only possible answer is, the people who *believe*, at least, that they themselves, or certain other people, are *not* thoroughly selfish. But this belief is false in every case if the selfish theory is true, and could only arise, concerning ourselves, from self-deception, or concerning others, from successful hypocrisy on the part of those others. Yet hypocrisy and self-deception are among the very things which the selfish theory is intended to elicit condemnation of. So there is and must be, if the selfish theory is true, literally no one at all in the gallery who could feel surprise or moral indignation at the revelation of universal selfishness.

A more familiar form of the same objection to the selfish theory is this: if humans had been uniformly and invariably selfish to begin with, how could they ever have come (as they plainly have come) to regard selfishness with moral disapproval? The problem is actually more general than that. How could they ever have come to regard *anything* with moral disapproval; or for that matter, with moral approval either? Still, it is clearly the special case, of the moral disapproval of *selfishness*, which poses this general problem in its sharpest form to the selfish theorist.

The Enlightenment had an answer (of a sort) to this question, along predictably superficial lines. Mandeville is a representative example of it. Morality in general, he says, and the moral disapproval of selfishness in particular, came into existence through "the skilful management of wary politicians."[21] Human beings were originally, just like all other animals, entirely selfish, and devoid of any ideas of morality. But, by working on their pride and shame, "law-givers and other wise men"[22] managed to flatter and dupe them into a taste for self-denial, altruism, and the moral point of view. In Mandeville's famous phrase, "the moral virtues are the political offspring which flattery begot upon pride."[23]

Have you ever heard of anything more ridiculous than this? It puts even Hobbes's and Huxley's "war of all against all" in the shade, and could fairly be called the *Cave Man's* Cave Man way out of Darwinism's dilemma. It manages to be both a self-contradictory and a miraculous story.

Pride and shame, as everyone knows, are two of the emotions which most often *deter* people from acting selfishly. Yet Mandeville attributes these emotions to his supposedly selfish original men! Then, pride and shame are also emotions which, quite obviously, include a moral element. If, for example, you refrain from doing a certain thing because you would be ashamed to do it, then you think that thing *wrong*. A person who is influenced by either shame or pride cannot, therefore—logically cannot—be a stranger to all moral ideas. Yet Mandeville's original men are supposed to be such strangers.

But if human being *had* been at first devoid of any ideas of morality, how could even the most "artful" among them—how could anyone, short of a demigod—ever have got them, not merely to act on moral ideas, but to *understand* moral ideas in the first place? With writers like Mandeville in mind, Hume justly said: "Had nature made no such distinction, founded on the original constitution of the mind, the words *honorable* and *shameful*, *lovely* and *odious*, *noble* and *despicable*, had never had place in any language; nor could politicians, had they invented those terms, ever been able to render them intelligible, or make them convey any idea to the audience."[24]

Even if this miraculous imposition of morality, on selfish and pre-moral human animals, could be explained, it would still leave unexplained the earlier and greater miracle: how the "law-givers and wise men" *themselves* came into possession of moral ideas.

But we are here clearly engaged in breaking a butterfly on a wheel. As Hume also rightly said, "nothing can be more superficial"[25] than the Enlightenment's attempt, supposing humans to be originally selfish and non-moral, to explain how they came to be otherwise. It is entirely characteristic of the Enlightenment, of course, to ascribe human morality and altruism to an external cause, and in particular to an *educational* one (in a broad sense). But it does not make any *biological* sense at all.

IN RECENT DECADES selfish theorists—neo-Darwinian ones this time—have returned in force to this old question of the origin of altruism. With them, of course, this question takes the new form of how altruism could have evolved, and not been eliminated by natural selection. A vast amount of mathematical and computing ingenuity has lately been expended on this question, by socio-biological and other games theorists, and especially by prisoner's dilemma monomaniacs.[26] Unfortunately, however, no one knows how much relevance the results of all this effort have, to any *actual* events in biological history; or even whether they have any such relevance at all.

I cannot, for my own part, take any interest in these neo-Darwinian "solutions to the problem of altruism." The reason is that I cannot take "the problem of altruism" seriously in the first place. Neither should anyone else take it seriously. For if you do—if you start off by thinking that altruism is a problem—then logically you will not be able to stop short of thinking (with Richard Dawkins) that it is a *problem* why mothers do not welcome baby snatchers, or of thinking (with E. O. Wilson) that it is a *problem* why he does not always try to kill or maim the professors who are his enemies. But it is perfectly obvious that once Darwinian armor-plating has reached *this* degree of thickness, it is completely impenetrable by common sense, or even sanity. The fact is, there is *no* problem about human altruism. The only problem is Darwinism and neo-Darwinism.

I do not believe for one moment that human beings ever *were* devoid of altruism; and my reasons are the ones I gave in Essay I for ridiculing Huxley, when he says that humans were once engaged in a Hobbesian war of all against all. I do not believe for one moment that human beings ever were regularly subjected to a struggle for life, and a child mortality, anything like as severe as Darwin and Malthus imply that they must always be; and my reasons are the ones I have given in Essays 2, 3, and 5, for thinking that Darwin and Malthus were badly wrong about man from the very start. As for those sociobiologists who by implication deny the very existence of human altruism: my reason for disagreeing with them is simply that I am not a lunatic.

There is no reason whatever, apart from the Darwinian theory of evolution, to believe that there ever was in our species an "evolution of altruism" out of a selfish "state of nature." People believe there was, only because they accept Darwin's theory, which says that there is always a struggle for life among conspecifics, whereas there is no such struggle observable among us now, but a great deal of observable altruism instead. The right conclusion to draw, of course, is that Darwin's theory is false. But the conclusion usually drawn is the Cave Man one: that there must have been an evolution—admittedly difficult to explain—from an originally selfish human nature into our present altruistic and tax-paying state. This, however (as a great philosopher said in another connection), is first raising a dust, and then complaining that one cannot see.

There is extremely little observable struggle for life, and much observable altruism, among present-day kookaburras, too. They are loyal spouses, and devoted parents to their slow maturing young; and if you expect to see kookaburras squabbling over food like gulls, you should come prepared for a very long wait. Are we therefore to infer, because Darwin's theory of evolution is such a wonderful idea, that there must long ago have been a Hobbesian war of all kookaburras against all kookaburras? Are we obliged to generate a vast game-theoretical literature, to explain the mystery—a mystery indeed—of how originally selfish kookaburras could have evolved into the petit bourgeois kookaburras we see now? Presumably not. And yet if not, then why expend time and thought on similar stories, which are absurd as well as unnecessary, about the evolution of our own species?

But if, on the other hand, your faith in Darwinism is so profound that you simply *must* have human beings, not only in the remote past but now too, always engaged in a struggle for life so severe that it leaves no room for altruism and exacts a child mortality of 80 percent or more: well, if you have made that uncomfortable bed, you will just have to lie in it. And one of its minor discomforts is this: that you will have to reconcile yourself to performing, all your life, that evasive trick of which Hume rightly complained. That is, of calling certain facts—namely the facts of human altruism—a "problem" or a "difficulty" for your theory, when anyone not utterly blinded by Darwinism can see that these facts are actually a *demonstration of the falsity* of your theory.

THE SELFISH THEORY of human nature was always explicitly intended by its adherents to explode the belief, assiduously cultivated by priests and other obscurantists, that a vast gulf separates our species from all other animals. It was intended, as Darwinism was always intended, to bridge the gap between man and the animals, to mortify human self-importance, and to "cut us down to size." Now isn't that just too bad? Because a vast gulf *does* separate us from all other animals, in point of altruism, as in point of intelligence. That is simply a fact, and a very obvious one, even if it *has* been stated by a billion obscurantists.

V

A selfish theorist, especially a new-Darwinian one, is unlikely to have been swayed by anything I have so far said in this essay. I have not done anything against the selfish theory, and cannot do anything, except draw upon everyday experience and common sense. I have on my side nothing which could possibly be dignified with the name of "a scientific theory," let alone a successful scientific theory. Whereas neo-Darwinian selfish theorists have on their side the most successful scientific theory in the history of biology.

But even if we leave out of account everything to do with Darwinism, the selfish theory of human nature appears to have a certain advantage over my (or anyone's) denial of it. For selfish theorists have always been able to point to certain *experiments* which support their theory, and which appear to do so decisively. And everyone agrees that decisive experiments count for more than either everyday experience, or the support afforded by a successful scientific theory.

The experiments I mean are not of the ordinary kind, which are brought about by human contrivance. They are some of those "natural experiments" which happen without our doing anything to bring them about, but which are nevertheless *as though* we had arranged them in order to learn something from them. Solar eclipses are an example of natural experiments. We cannot diminish and then increase the energy which the sun's surface releases towards us, but during an eclipse of the sun it is in certain respects as though we could; and as a result, a good deal of solar physics has been learnt from eclipses, which could not have been learnt in any other way. Another example of a natural experiment is the people who are born blind but are otherwise normal. For they afford us a unique opportunity to estimate how much of ordinary educability depends on being able to see.

The natural experiments which seem to speak decisively for the selfish theory are all those instances in which people, subjected to extreme suffering, from hunger or torture for example, have been found to behave with regard to nothing except their own survival. The survivors of a shipwreck, for example, threatened with starvation, have sometimes killed and eaten one of their number. To avoid renewed torture, a political prisoner will often sentence to death his own associates still at large, by disclosing their identify to the torturers. And so on.

No one will deny the reality of such cases. And we seem to see, in such cases, every shred of civilization, morality, sociability, and even of the closest attachments of blood, peeled away, and human beings revealed as starkly selfish. Well, I do not deny that such things *are* evidence in favor of the selfish theory. But I do deny that they are *decisive* in its favor.

One reason is that these natural experiments on human selfishness do not all turn out the same way. It sometimes happens that among the survivors of a shipwreck or plane crash, all of whom are threatened with death from hunger or thirst, some bring water or food to others for as long as they are physically capable of doing so. Malthus and Darwin have been far too successful in suggesting that, in all animals, the impulse to sustain life by eating is sovereign and uncontrollable. Yet it is well known that self-starvation has been deliberately adopted as a method of suicide in many ages and countries. Then there are the familiar facts of fasting, on religious or on medical grounds; of dieting, on aesthetic or moral or medical grounds; and of the religious proscription of certain foods. However it may be with other animals then, it is absurd to suppose that human beings are under a biological necessity always to eat what food they can get. Yet the "shipwreck" and similar experiments are often described in such a way as to imply just that absurdity.

Then, political prisoners, repeatedly tortured to make them reveal the identity of their associates still at large, have sometimes defied their torturers and carried their secrets to the grave. Others, mistrustful of their own ability to withstand torture, have often found a way of taking their own lives before the testimony they were determined not to give could be extorted from them. North American Indian warriors, captured by members of another tribe, and subjected to the most terrible cruelties which human ingenuity could devise, have sometimes been known to die scorning their tormentors' efforts, and urging them to try harder.

I do not say that such cases are common, or even as common as the cases of the opposite kind, in which people under torture save their own lives at all costs, and keep their own suffering to a minimum. But they are perfectly real cases, and are quite as historically certain as the cases of the opposite kind. They may be incomprehensible cases, and indeed they are incomprehensible, to sociobiologists and to selfish theorists in general, as well as to most of the representative fashionables of the present day. (You would have to be *extremely* uneasy if, for example, your life depended upon a sociobiologist or a draft dodger standing firm under torture!) But then, as we have seen earlier, a great many of the most obvious facts of human life are incomprehensible to these classes of people: the fact, for example, that most mothers do not welcome baby snatchers.

No theory can be *established*, as distinct from being confirmed, just by the cases it successfully explains. It does not matter how many of these cases there are, or how well the theory explains them. For there may be other cases which the theory not only does not explain, but in which the observable facts are the very opposite of what the theory leads one to expect. That is how things stand with the selfish theory so far. If, therefore, a woman and her infant are both nearly dead from hunger, and she eats all the food there is and lets the child die, or if a tortured man saves his own life at the cost of the lives of three or four associates, then the selfish theory is indeed to some extent confirmed by these natural experiments. But they are not *decisive* in its favor, since there are many other cases which tell *against* the theory.

There is another reason why such cases are not decisive. This is that they are and must be *exceptional* cases. They might be common enough, indeed, in absolute numbers. Perhaps, around the globe, it happens many times each day that a woman can survive only by letting her infant die, or that a man can survive only by condemning three or four others to death. Still, such cases could not be usual, or make up most of human life, for a very obvious reason. If a woman could *usually* prolong her own life only at the cost of her infant's life, or if a man could *usually* prolong his life only by three or four other men dying, there would very soon be no one left.

If such cases could not possibly be usual, then still less could they be universal. And yet, what is it that these exceptional and gruesome cases are cases of? Why, just of *a struggle for life among conspecifics*. Though Darwin has made us all so extremely familiar with the idea of a struggle for life among conspecifics, you probably did not at first recognize my starving woman, or my tortured man, *as* cases of that struggle. Still, that is what they *are* cases of, whether you at once recognized them as such or not. And Darwin said—let us remember—that a struggle for life among conspecifics is universal and perpetual in every species!

ACCORDING TO selfish theorists, these cases, in which people condemn others to death in order to save their own lives, are exceptional only in that the veil or veneer which usually disguises human selfishness is for a moment stripped away. The selfish theorist identifies this veil or veneer with the demands for self-restraint, unselfishness, and cooperation which every human society imposes upon its members. But below this veneer, the selfish theorist says, human beings, even the most highly civilized ones, are really just as selfish as savages, sharks, or wolves, and will always reveal themselves as such when circumstances, such as torture or starvation, remove all the pretences and the superficial amenities of ordinary social life.

This idea, that civilization, morality, unselfishness and self-restraint, are only superficial and misleading appearances, disguising our selfish, savage, animal nature, I will call for short "the veneer idea." It is one of the darling ideas of the modern Enlightenment. It first surfaced in the Greek Enlightenment of the fifth century, but then almost vanished for nearly 2,000 years, presumably because of its manifest incompatibility with Christian beliefs about man. Since about 1700 A.D., however, it has triumphed almost everywhere, and exercised an enormous influence on the minds of all Enlightened persons.

There are several main variants of the veneer idea. The Freudian variant is the most widely diffused one. According to this there is, in the mind of every civilized person, an "id," which is the assemblage of all the impulses natural to us as animals, and a "super ego" which is the inner representative of the demands of society for unselfishness, order, cooperation and decency. In the Freudian variant, to act as a kind of gasket or washer between these two layers of the kind, an "ego" is also postulated. The ego has the thankless and endless task of putting together, from moment to moment, a workable compromise between the brutish impulses coming to it from below, and the unrealistic moral expectations being pressed upon it from above.

The veneer idea has always been especially congenial to those Enlightened people who have the cause of sexual emancipation most at heart. The Freudians are, of course, the most famous and effective sub-class of this class, but they were very far from being the first. Sexual emancipation was from the start high on the agenda of all Enlightened people, being judged in the eighteenth century only a little less urgent than the destruction of religion and the overthrow of monarchy; and the veneer idea was early put to work for this worthy cause. Diderot, for example, put it to work in two famous books. One was *The Nun* (published in 1796 but in circulation long before that), which is a proto-Freudian cautionary tale, about the awful effects of repressing sexual impulses. The other was his imaginary *Supplement* (published in 1796 but again in circulation long before), to the *Voyages* of the explorer Bougainville. The burden of Diderot's *Supplement* is simply this. "How much better they arrange sexual matters in Tahiti! How much happier would we Europeans be, if *our* sexual arrangements were not poisoned by priestcraft, and impeded by a thousand imaginary obligations about fidelity, parenthood, and property!"

Margaret Mead, of course, made the same exciting "discovery" all over again, in Samoa in the 1920s. In return, a grateful world made her into an oracle, both scientific and moral: presiding over American anthropology, advising American parents, from the President down, on how to handle the turbulence of their adolescent children, etc. It was not until more than fifty years later that her account of Samoan sexual life was proved to be entirely false:[27] an optical illusion produced by the Jazz Age spectacles through which she had looked at her subject. Diderot had at least known that his *Supplement* to Bougainville was a work of fiction.

Alas, the falsity of what these and other Enlightenment missionaries wrote about the South Pacific did not make the damage they did to the fabric of human society, both at home and abroad, any the less great. The Enlightenment dream of sexual salvation under the palm trees, is, to this day, one of the things which hourly brings down yet more gigantic plane loads of tourists upon those most Unfortunate Isles. About the blessings which we owe at home to such beacons of sexual Enlightenment as Diderot, Freud, and Mead, it can hardly be necessary to speak. Just read the newspapers, or reflect on the marital and parental careers of yourself and your friends.

The other main variant of the veneer idea is the Darwinian and neo-Darwinian one. That Darwinism should *be* a variant of the veneer idea was inevitable from the start. For the Darwinian theory said all along that a struggle for life is always going on among the members of every species; and yet there is *no observable* struggle for life in any human society, savage or civilized. (Except, again, in such exceptional cases as those I have spoken of, of a woman having to choose between letting herself and letting her infant die of hunger, or a tortured man obliged to choose between condemning himself and condemning his associates to death.) So, if you intend to stick to the Darwinian theory, you simply *have* to say that, in the human case, most of the time, the struggle for life is going on *below the surface of society*: concealed by the veneer of unselfishness, considerateness, and so forth.

The denizens of the dark but "real" underworld of human life are conceived somewhat differently by Freudianism and Darwinism. In both, of course, they are thought of as being selfish, anarchic, and non-moral; but their selfishness is focused on rather different objects in the two cases. In Darwin's "struggle for life" the competitors, whether human, animal, or vegetable, are bent on nothing but survival and reproduction: that is, on food and future offspring. The impulses which make up Freud's id, on the other hand, care nothing about future offspring, and indeed lack the very concept of the future. More strikingly, they show no interest in food: surely an amazing indifference, this, in impulses which belong to the animal basement of human life. I suppose the explanation must be that id impulses benefited, like everyone else, from the marked improvement in economic conditions which occurred between Darwin's time and Freud's. So what *have* id impulses got on their minds, if not food or offspring? Why, just sex. In short, they are really just civilized people of the twentieth century. In no earlier age could Freud's theory possibly have been taken seriously. Imagine trying to convince Isaac Newton, or John Knox, or Eric the Red, or Charlemagne, or St. Paul, or Aristotle, or an élite and exclusively homosexual unit of the Spartan army in 450 B.C., that their only *real* goal in life is—to copulate with women!

Despite these differences of detail, it is obvious enough that Darwinism and Freudianism are only variations on a common theme, and what that theme is. It is that such things as self-restraint, cooperation, and consideration for others are merely part of a thin disguise which society places over our selfish and non-moral animal nature.

DESPITE THE WIDESPREAD and longstanding acceptance that the veneer idea has enjoyed, and still enjoys, it is false, and even obviously false. For it compels us to ask a certain simple question, and yet cannot answer it: namely, *whence the veneer?* What could have brought such a thing into existence in the first place, or kept it in place if it *had* once come into existence?

We saw earlier, in Mandeville, a clear case of this insoluble difficulty. Men were originally, and are still by nature, he says, selfish and non-moral; but then morality was brought in by "law-givers and wise men." Yet law-givers and wise men are just men, after all. Hence if what Mandeville had first said about human nature were true, they could not possibly be unselfish or considerate or cooperative themselves, and even if they were, could not possibly have interested their savage or wolfish fellows in acquiring these unheard-of and plainly deleterious attributes.

Freudianism suffers equally from the same fatal defect. According to this theory, all the impulses which are natural to human beings are collected in the id. But then, to "drive" the super ego, what source of psychic energy is left? The super ego has the onerous task of imposing a veneer of regard for others, order, and decency, on the selfish, anarchic and indecent impulses of the id. It is supposed to be a fearfully effective engine for this work too, inflicting untold misery on nuns and countless others, by denying satisfaction to their natural sexual impulses. But what fuel can this mighty engine of repression possibly be running on?

It does no good to answer, as Freud in effect did, "fathers," or "fathers as perceived by their sons." Like Mandeville's "law-givers and wise men," fathers are just certain men, and are in any case sons themselves. Hence if the Freudian theory is true, all *their* natural impulses are id impulses too: that is, selfish, anarchic, and non-moral. Thus we get no nearer, by going back a generation, to explaining the creation of the veneer of morality. How on earth *was* that veneer ever brought into existence? It must *be* on earth, of course, for Freudians: they are far too Enlightened to postulate the intervention of a God or demigod. Yet nothing in *human* nature, as Freudians portray it, could possibly have given rise to self-restraint, cooperation, and the moral point of view.

Darwinism has been dogged, from 1859 to the present day, by essentially the same problem. Not exactly the same one. With Freudianism, and with Mandeville, the problem is how morality and altruism could ever have come to exist. With Darwinism the problem is rather how those things (however they might have come to exist) could have *survived*. For Darwinian theory says that there is always a struggle for life going on among the members of every species. So why was not every tender shoot of altruism or morality always promptly sheared off by natural selection?

If you on an impulse make an altruistic "offer" to some of your non-altruistic conspecifics, they will—if words mean anything—close with your offer, and thereby improve their own chances of surviving and reproducing; but not yours. If you make a habit of this kind of thing, there is only one way matters can end for you, and for any offspring you may manage to leave who inherit your amiable disposition. Your lineage, far from becoming one of "the favored races in the struggle for life," will quickly be extinguished.

This is the form that the question "whence the veneer?" takes in the case of Darwinism. If the members of every species are always engaged in a struggle for life with one another, and if human beings were selfish and non-moral animals at first, how could even the least little bit of morality or of altruism have escaped being eliminated by natural selection?

If it is an insoluble problem for Darwinism how even the least bit of altruism could survive, then think of the scale of the problem for Darwinism which is presented by the advanced societies of the present day. In these, altruism has not only survived, but spread like wildfire, and even assumed monstrous proportions. In fact it long ago reached a stage of morbid gigantism which Malthus, as an economist, had warned against, as tending to the destruction, both of all existing wealth, and of the kind of person who can create wealth. Think of our stupendous present expenditure of money and effort on public health, education, unemployment relief, and the rest. (Quite a veneer, this!) And then recall that there are thousands of Darwinians who are, at this very moment, puzzling their heads as to whether there is such a thing as altruism, with most of them gravely concluding, faithfully to their theory, that there *isn't*, really! If only one had the power of language that would be needed to do justice to this colossal "scientific" farce!

Why *do* we devote all this money and effort to paying many unemployed people who refuse to work, to "educating" many who are incapable of being educated, to caring for the health of many who are hopelessly moribund or else healthy? Why *do* selfish girls, and all the rest of us, pay taxes?

I hope you will not say, "Because we know that otherwise we will be fined or imprisoned"; or, what comes to the same thing, "Because behind the tax man stands the Sword." That is, of course, the Darwinian answer, the answer of other selfish theorists such as Hobbes and Mandeville, and the answer of Enlightened people in general. But it is a hopelessly superficial answer, in precisely the Enlightenment way.

There is nothing whatever special about the people who make and enforce the tax laws. You could not find a more ordinary bunch if you tried. The tax men, like Mandeville's "wise men" and Freud's "fathers," are just some more people. They are not, in their own persons, any more altruistic than other people, and it would not help them get their hands on our money if they were. They are even, if you can believe Darwin, *competing* with you and me and everyone else, to survive and reproduce. They are certainly not the only group of people who tell us all the time that we should send our money to them. Cranks of a thousand kinds are always doing that, as well as assorted criminals. Severe penalties for any disregard of their directives would also be enforced by the cranks and criminals, if they could. Some of the criminals, of course, can and do enforce such penalties.

It cannot be due to any biologically irresistible selfishness on our part, then, if most cranks and criminals do not get our money: since the tax men *do* get it. *Of course* they have the power of the Sword behind them, and no one else, or near enough no one else, has. But the right questions to ask are, *why* it is in *their* undistinguished hands that the power of the Sword has come to be placed, and why it is that that power is put, on the scale that it is, to such unswordlike uses as helping the sick, the ignorant young, and the unemployed poor? No answer is possible, except that the tax men are known to execute, on our behalf (however erratically), our altruistic determination to help those people.

The same conclusion is forced on us independently, and in an unexpected way, by those parts of the world—western Sicily, and parts of America and Australia—where organized crime constitutes a "parallel government," and possesses its own power of the Sword. The *Mafiosi*, it must be admitted, are less concerned than official governments are with either education or unemployment relief. They consider that they themselves are able to provide whatever education and employment is really needed. But as to hospitals and public health, they put all official governments to shame. Their maternity wards in particular, and their provision for the care of the aged—based entirely on private generosity—are always far better than those in districts where the whole burden is carried by the *public* purse. The best equipped maternity section of any Australian hospital is that of Griffith Base Hospital.

Since selfish theorists must explain, and yet cannot explain, the veneer of unselfishness in human life, it is not surprising if they contradict themselves when they describe the pre-veneer men whom they postulate. Mandeville, as we saw, said that they feel shame and pride, but have no conception of morality: which is just like saying of someone that he loves the fragrance of roses but has no sense of smell. And selfish theorists fare no better in their descriptions when they discover, or think they have discovered, a society of *post*-veneer people. Colin Turnbull's book *The Mountain People*[28] is the report of a recent supposed discovery of this kind.

The book concerns a central African people, the Ik, who, by the time Turnbull found them, had been expelled by various pressures from the hunting grounds which had previously sustained them, and were obliged to try to live by farming land which was simply too dry and poor to yield them a living. They were dying out fast from mere hunger, and the survivors were in a dreadfully reduced physical condition: a claim amply supported not only by Turnbull's text but by his numerous photographs.

As a result, Turnbull tells us, the veneer which usually disguises human selfishness has quite fallen away among the Ik. They have "dispensed with the myth of altruism" (page 130). All social bonds, and even almost all bonds of kindred, are things of the past with them. Here, the structure of society has been resolved back into its original constituent atoms: hungry animals. Selfishness reigns, unchecked and undisguised, and is focused on just one object: food. That is the only good of which the Ik any longer have any conception. Indeed, their "very word for 'good,' *marang*, is defined in terms of food. 'Goodness,' *marangik*, is defined simply as 'food,' or, if you press, this will be clarified as 'the possession of food,' and still further clarified as '*individual* possession of food.' Then if you try the word as an adjective and attempt to discover what their concept is of a 'good man, *iakw anamarang*, hoping the answer will be that a good man is a man who helps you fill your own stomach, you get the truly Icien answer: a good man is one who *has* a full stomach" (page 112).

The Mountain People became a best seller. It could hardly have failed to do so, since it has the same irresistible ingredients as such earlier best sellers in the same vein as Mandeville's *Fable of the Bees*, and Hobbes's *Leviathan*. Civilized people love to shudder at shocking stories about "savage" life, and to think "How unlike *our* situation, thank God!" But they also love to think, if they are Enlightened, "Apart from our thin veneer of civilization, how exactly *like* these people we are!" Turnbull's book had the additional advantage that, being published in 1972, it was perfectly timed to catch the neo-Darwinian, and more specifically, the sociobiological vote.

BUT IT ALSO resembles Mandeville's book, in that it wears falsity on its very face, and indeed on almost every page. More specifically, it wears *self-contradiction* on almost every page. The following are a few examples of things which Turnbull tells us, that are simply incompatible with the general picture he himself paints, of all social ties having been dissolved. I could easily give a hundred such examples.

The favorite son of a man named Lomeja died one night in the family house. "Losealim [Lomeja's wife] had suggested burying the body next morning. Lomeja had said No, better bury it in the compound right away while it was dark, *otherwise it would involve a funeral and*, of all things, *a feast*. The boy was not worth it, he was only a boy. Losealim refused, so Lomeja beat her, and it was she whom I had heard crying, because she had been so badly beaten and, on top of it, made to dig the hole. And Lomeja was looking stricken because now everyone knew that [his favorite son] had died, and *they would expect him to give a feast*" (page 108, italics added). I suppose that some other Ik men, though all of them were on the brink of starvation, were financially ruined because they were expected to give a large dowry to their daughters, or to pay for their wives' extravagant credit card shopping.

Though there is no actual shopping in Turnbull's book, there are dozens of references to buying and selling, both among the Ik and between them and neighboring tribes. Theft is common and increasing, but is also acknowledged to be wrong (see page 143, for example). But buying and selling, and theft recognized as wrong, imply *the existence of rights of property*. And how can there be rights of property, where no one any longer has any conception of good, except food?

Not only are there rights of property among the Ik: there is even a police station! Turnbull implies, indeed (page 205), that the police are not altogether honest, or able to be trusted with the custody of young women. But that, of course, is no more than can be said with truth of the police in many far happier lands. Yet it does not elsewhere prevent them performing some of the functions for which police exist, such as the protection of life and property; and Turnbull says nothing to suggest that these police do not also perform those functions to some extent. But what is a functioning police station doing in the middle of a Huxleian-Hobbesian competition for food, or a Darwinian "struggle for life"?! The conjunction simply does not make sense.

Even the strongest *biological* ties, Turnbull would have us believe, are vestigial or vanished among the Ik. The "mother throws her child out at three years old. She has breast-fed it, with some ill-humor, and cared for it in some manner for three whole years, and now it is ready to make its own way. I imagine the child must be rather relieved to be thrown out, for in the process of being cared for he or she is carried about in a hide sling wherever the mother goes, and since the mother is not strong herself this is done grudgingly. Whenever the mother finds a spot in which to gather, or if she is at a water-hole or in the fields, she loosens the sling and lets the baby to the ground none too slowly, and of course laughs if it is hurt. I have seen . . . this many a time. Then she goes about her business, leaving the child there, almost hoping that some predator will come along and carry it off. This happened once while I was there, . . . and the mother was delighted" (page 113).

That is, Ik mothers carry and care for their babies for three years, *and* during that time are pleased if they get hurt, or are killed by a predator! Could there be a more glaring inconsistency? It is so glaring, in fact, that it actually embarrasses the reader. Turnbull was plainly so determined to paint the blackest picture possible, that he did not mind, or did not notice, when he openly contradicted himself; which naturally makes the reader wonder whether the book was not dictated by some *private* misery, which the author had merely "projected" onto his nominal subject matter.

If Turnbull had been content to describe how protracted hunger had *lessened* the strength of social and even biological ties among the Ik, he might no doubt have written a truthful book. But of course he would not have written a best seller that way. Nor would he have convinced his readers that every bit of unselfishness in human life is only a polite polish which people put on when the living is easy: which is what he wanted to convince his readers of. He did convince many of them, too. One of the admirers of *The Mountain People* is the sociobiologist Richard Dawkins, who refers to it as though it were an authoritative scientific work.[29] Well, he would, wouldn't he? Not surprisingly, another one of Dawkins' anthropological "authorities" is Margaret Mead.[30] Deep calls to deep.

IT IS A FATAL *internal* disorder of the veneer idea, that it makes it impossible to explain the existence of the veneer. But there are, in addition, external facts which are equally fatal to the veneer idea.

The idea is, that people like you and me, underneath our surface civility, etc., are really just savages or wild animals after all. But unfortunately for the selfish theory, it has turned out that "savages" *themselves* are not savage, and that "the beasts," or "the wild animals," are neither beastly nor wild. The last hundred years of ethology and anthropology, if they have established anything at all, have established this fact. The difference between us and "savages," whatever it may consist in, certainly does not consist in their being overtly selfish, anarchic, and non-moral while we are covertly so. And the difference between humans and animals, whatever it may consist in, certainly does not consist in their being wild or beastly, while humans are not.

Ever since ancient times it has been a favorite saying of selfish theorists that *homo lupus homini*: man is a wolf to man. But the wolf, the supposed paradigm of wildness and beastliness, has let the selfish side down completely. For on closer study it has turned out that you would have to go a very long way to find animals more assiduous than wolves in the discharge of all their parental, domestic, and social obligations, or less prone to anti-social outbreaks of individual self-seeking. If their devotion to a meat diet is beastly, so is ours. Their feeding habits are not, any more than are those of any of the higher mammals, a mere competition to engulf the most food. In what, then *does* the "wildness" of wolves consist? In their being dangerous to man, if they can get away with it? But so are we to them, if we can get away with it. And so the comparison goes on, with no more wildness or beastliness appearing on one side than on the other.

If a selfish theorist were to arise among wolves—a vulpine

Mandeville or Hobbes, say—he would undoubtedly tell his conspecifics that *lupus homo lupo*: wolves are men to other wolves. He would be profoundly wrong about his own species, of course, and would be gratuitously insulting them, too. His conspecifics would tell him so, if they had any sense. But he would be no *more* wrong, and no more gratuitously insulting, than *our* selfish theorists are about our species.

It has turned out, in fact, that no animals are "wild," in the sense that we meant when we first ignorantly called some of them so. That is, in the sense that their behavior is free from social constraints, and is dictated entirely by the self-interest of each individual. But this is also the sense in which selfish theorists believe that *we* are, below the veneer, wild animals. Hence, in the sense in which they think we are wild animals, *there are no wild animals at all*, either of our species or of any other.

There is, indeed, in horses and a few other species, a process—a profoundly mysterious process—of being "tamed" or "broken" or "broken in." That is, a process of being made thoroughly amenable to a system of discipline which humans wish to impose on them. But the state of these animals before they are broken in, and the state to which they revert if they afterwards return (as we say) to "the wild," is simply a *different* state of social discipline: one imposed on them by their own conspecifics. It is not a state of anarchic indiscipline and individual self-seeking. To think that it is, is exactly like the mistake of children and fools, that on a pirate ship discipline is non-existent, or less severe and arbitrary than it is on a legitimate ship. In fact, of course, it is more severe and arbitrary. Well, it had *better* be, hadn't it?

Likewise, "savages" do not differ from civilized people by having a thinner veil, or a lighter veil, or no veil at all, placed over the selfish animal nature which is common to them and us. In fact, as anthropologists have established a hundred times over, the boot is altogether on the other foot. More than any other humans, civilized ones go or stay, do or don't do, decide *pro* or *con*, as their individual inclinations and beliefs prompt them.

"Savages," by contrast, have their behavior rigorously prescribed for them, at almost every moment of life, by gods or ancestors or elders or priests or chiefs, or at any rate by *some* external authority. And the more "savage" a tribe is, the more comprehensive and vice-like is the grip of social prescription on the lives of its members.

This is by now generally known. But it has not been generally noticed that it is fatal to a favorite idea of Freud. He told his readers that civilized people are discontented, because of the uniquely heavy burden of "instinctual renunciation" which civilization, as distinct from "savage" life, imposes. No wonder his civilized readers were more than willing to believe him, especially since he implied that we could probably by now well afford to loosen the clamps just a little bit! But in fact the *mores* of any "savage" society require far more "instinctual renunciation" than civilization does. Enlightened opinions, as I said earlier, are always superficial, and the causes of our civilization's discontents lie a good deal deeper than a Freud could imagine. In any case, we have now actually tried, on a gigantic scale, the remedy for discontent which Dr. Freud recommended; but alas, we are more discontented than ever.

It is not opening the way to anthropomorphism to say that "the beasts" are not beastly. It does not mean, for example, that (as Hilaire Belloc sang) a tiger

> Is kittenish and mild,
> And makes a splendid playfellow
> For any little child.

Nor, in saying that "savages" are not savage, is there any concession to a fatuous and dangerous "multiculturalism." It does not mean, for example, that some Dayak head hunters of nineteenth-century Borneo would fit in well in an Australian suburb of the present day, or would do so if it were not for "racism." The differences between humans and other animals, and between civilized humans and others, are exactly as great and as many as—as they are found in fact to be! And they are found to be, in fact, many, vast, and intractable.

Human societies are almost inexhaustibly various, but there is one thing which *no* human (or even animal) society is even remotely like: namely, "savage" life, and civilized life below the veneer, *as selfish theorists conceive it.* They think of people as though they were the molecules of a confined volume of gas, which have no mutual sympathy, or any other influence, except by way of *collisions* with one another. This is the selfish theory to a T, as long as you impute to each molecule a ceaseless and exclusive regard to its own interests. The only thing wrong with this idea is that there is nothing whatever in reality which corresponds to it.

AND YET the selfish theory, as I said at the beginning of this essay, never dies. Indeed, it somehow always retains its irrepressible appeal, even after the weightiest blows of criticism. In fact its influence is never entirely extinguished, even in the minds of those who criticize it most severely and justly.

Part of the reason is that all of us inwardly know something of the great extent of our own selfishness, and reasonably suspect that it is much more extensive than we know. But the main reason is that we can never forget for long those shocking "natural experiments" on human nature, of which I spoke earlier. That is, the survivors of a shipwreck resorting to murder and cannibalism, the tortured prisoner who at last names his accomplices, the starving mother who feeds herself and lets her baby die, and so on.

It is this kind of evidence which every selfish theorist will always fall back on in the end, and which seems, indeed, to clinch the question in his favor. It is also evidence which always comes back to haunt the critic of the selfish theory, even after he has done his very best to demolish that theory. And these horrible old stories bring the veneer idea back with them. For they seem to admit of no other explanation than that everything in our civilization, morality, or culture which points to unselfishness, is only

159

a superficial and misleading appearance. This picture "holds us captive." But I will make one last attempt to release us from our captivity to it.

Suppose you caged a whole family of kookaburras—mother, father, and several half-fledged babies—and slowly starved them, until only the mother and one baby were left alive. If at this point you put just enough food in the cage to keep one of the two alive, it is likely enough that the mother would eat it all and let the baby die. But it is absolutely certain that, well before that time, the manners and morals of the two adult birds, in relation to food, would have changed considerably for the worse. It is highly probable that they would even have had to revise their ideas about what *is* food, and what is not: whether, for example, a dead fledgling of their own is food or not. (The babies, since they had no manners or morals to lose in the first place, either in relation to food or to anything else, would probably not have changed *their* deportment or ideas.)

Now, is this experiment evidence—any evidence at all—that at some earlier stage of their evolution, kookaburras were all engaged in a ruthless struggle for life with their conspecifics: a war of every kookaburra against every other, in which "nice guys finished last"? Nowadays, of course, these birds lead quite humdrum lives, rather like yours and mine, in which domestic and parental cares bulk far larger than the exigencies of war, and in which fighting is exceptional. But aren't we seeing, in our starvation experiment, that the real kookaburra is, even now, the ancient *Hobbesian* kookaburra? Doesn't the experiment show, or at least suggest, that under the ordeal of starvation, the superficial veneer of present-day kookaburra civilization, or morality, or culture, is stripped away, and the old wild selfish kookaburra nature revealed?

No. No one would say such a stupid thing. The experiment does not support that hypothesis. In fact the hypothesis is so stupid that no experiment *could* support it. Kookaburras do not *have* a civilization, or morality, or culture, as distinct from certain products of their biological endowment. Therefore, whatever it is that is stripped away in the starvation experiment, it cannot possibly be *that*.

Suppose you keep a number of rats in a clear glass case, in which there is nothing that a rat could use to conceal itself. If you choose a particular rat and subject it to repeated torture, then very soon that rat, whenever it sees you coming, will—what? Try to hide behind or under some other rats, of course.

Every rat, it is safe to assume, is plentifully endowed with strong and purely selfish impulses, especially directed to getting food and getting mates: id type impulses. And yet in rat life, as a human life, there are countless appearances, at least, of unselfishness, cooperativeness, altruism, and so on. Since these things are plainly "against the grain" of the selfishness of individual rats, they must be ascribed to imperatives of rat civilization, morality, or culture, which would certainly have been internalized in the super ego of each rat. But aren't all the appearances of unselfishness, etc., in rat life, superficial and misleading? The rat super ego would certainly say, as the human one does, "Don't let others in for torture in order to avoid it yourself." But doesn't our experiment show that, under the ordeal of torture, the veneer of rat civilization is simply blown away, like the flimsy thing it is, by the reality of rat selfishness?

Again, no. In fact neither this experiment nor any other would be evidence for so ridiculous a hypothesis. Rats do not have any civilization, culture, or morality, as distinct from certain effects of their biological inheritance. So whatever it is that is stripped away from rats when they are tortured, it cannot possibly be that.

Yet a human prisoner, who names his accomplices in order to avoid further torture, behaves in essentially the same way as a rat prisoner that, in order to avoid further torture, tries to hide behind other rats. And a human family, if it were starved like the kookaburra family, would behave in essentially the same way. The adults would have their manners and morals, in relation to food, changed for the worse, and their ideas changed, too, about what is food and what is not. But the veneer idea is simply not available to us for explaining the behavior of starved kook-aburras or tortured rats. Its Darwinian variant, and its Freudian one, are equally out of the question. *No one* will believe in a veneer of kookaburra civilization, or in a rat super ego, or in their respective counterparts: an ancient race of Hobbesian kookaburras, now extinct, and a rat id, harboring all the low impulses that are frowned upon by a rat society.

But if the veneer idea is not even a possible explanation of what kookaburras or rats do under starvation or torture, it cannot be *the only possible* explanation of the essentially identical things that humans do under the same conditions. Of course we have, as rats and kookaburras do not, a culture, a morality, a civilization, which is not prescribed automatically by our biological endowment. But the rat and the kookaburra experiments should at least teach us, since their outcomes differ so little from those of similar experiments on humans, that we do not *have* to conclude, from the experiments, that our unselfishness is only an appearance, and our selfishness the only reality. There must (in other words) be *other* possible explanations, of what happens when a tortured man names his accomplices, or a starving woman feeds herself and lets her infant die.

I HAVE ANOTHER explanation to suggest. I do not dispute that, in these exceptional cases, *something* is stripped away from human character: something which normally is firmly in place. But the question is, what is it? It is certainly not mere politeness, or expressing more interest than one actually feels in the well-being of others. Such things abound in civilized life, of course, but they can hardly be what is in question here. After all, it does not take torture or starvation to strip away *that* sort of veneer! Many people drop all of that, every time they reenter their own homes.

The Enlightenment answer, of course, and hence both the Darwinian and the Freudian answer, is that what is stripped away under torture or starvation is the veneer: everything which does *not* belong to our biological nature, but has been added to us by nurture, or by education (taking those words in their broadest sense).

But I have given two reasons why this answer cannot be right. One is that kookaburras and rats have no veneer, and yet have *something* stripped away from them by starvation or torture, and then behave in a selfish way. The other is that the Enlightenment answer makes it an insoluble mystery why there is in our case a veneer at all. If all our natural impulses were id type ones, a super ego could never have got started; and no shoot of altruism could ever have survived natural selection, if humans had all been at first completely selfish and engaged in a struggle for life.

Enlightenment opinions about man are always hopelessly "external" or unbiological: they always exaggerate quite ridiculously the amount that education can achieve. They were, from the start, predestined to culminate in the utter biological blindness of a B. F. Skinner or a T. D. Lysenko. That paradigm of Enlightenment, William Godwin, had maintained in 1793 that, once we got education really right, not only religion and monarchy would be things of the past: so would aging, senescence, and death. No: "that way madness lies." The right answer to my question *must* lie on the other side: the side of biology, or of nature, not nurture.

The right answer will soon suggest itself, once we look for it on that side, if we bear in mind that we have to explain not only the behavior of humans under torture or starvation, but that of kookaburras and rats under the same circumstances. The clue is provided by those baby kookaburras which I mentioned earlier.

Starvation (as we saw) could not strip *them* of their morals and manners in relation to food, for the simple reason that they have none. Do you want to see animals that really *are* selfish, anarchic, and non-moral? Then you need go no further than a nest full of baby birds. *They* are your true "savages" or "wild animals"; at least, they come as close as anything does to realizing that Enlightenment myth. Just as a baby plant, in normal circumstances, embodies just one overpowering imperative— "Water, water!"—so does a baby bird, in normal circumstances, embody nothing but the imperative "Worms, worms!" And he does not mean "Worms for me *and* my sister here."

If a starving mother kookaburra with one surviving infant, or a starving human mother with one surviving infant, is given just enough food to keep one alive, she will almost certainly eat it all herself and let the infant die. Now, notice that in doing this, she does *what the infant would have done*, given the chance. The infant, whether human or kookaburra, would *certainly* have eaten all the food, with no compunction or even consciousness of the inevitable effect on its mother. A human mother, at least, might well feel, even *in extremis*, some compunction about the consequence for the infant, as she eats all the food. But the point of importance is, that both the human and the kookaburra mother, in doing the selfish thing, would be doing the *infantile* thing.

Surely we can now see our way out of the hopeless attempt, of the Enlightened in general and the sociobiologists in particular, to make out a "savage" or a "wild animal," thinly disguised, in every civilized man or woman. That is simply stupid: they are *not* that. But it is quite certain that they once *were* completely selfish, though also completely helpless, infants. Even now, if they are made sufficiently miserable and helpless again, by torture or starvation, they will revert—not to the "savage" or "wild animal" state but to their own state as infants. In other words, what is stripped away from us under starvation or torture is not cultural, but biological. It is not the successive layers of convention, education, morality, etc. It is the successive layers of biological development which are natural to our species between infancy and mature adulthood. That development is usually complete only when humans who are no longer selfish infants themselves find that they have offspring who are.

This suggestion, unlike the veneer idea, explains the behavior of rats and kookaburras under torture or starvation just as well, and in exactly the same way, as it explains the behavior of humans under the same conditions. Infant kookaburras and rats *are* wholly selfish: that is just a fact. Adult kookaburras and rats are not: and that too is just a fact. Adult rats, adult kookaburras, and adult humans whether civilized or "savage" simply do not want to engulf all the food that is going, as their infants want to.

They simply are not happy, as their infants would be, to transfer (if they could) any pain they meet with onto a neighboring conspecific. Those inclinations will come back to them again, of course, if their circumstances become taxing enough. But they are certainly not among their inclinations in normal circumstances.

And, again unlike the veneer idea, my suggestion does not make it an insoluble mystery why there is any such thing as culture, cooperation, or altruism at all. The strongest passion of our species (as I said much earlier in this essay) is for communicating with one another. That being so, it is no miracle if occasionally some gifted individuals, in their maturity, should invent something which makes for more communication than existed before, and which brings increased cooperation in its train. It might be a religious ritual, of marriage or of initiation into adulthood, say; a stirring war song; a system of signals between widely separated hunters; a tradition of expertise in handling large numbers of livestock; or many another thing. It would, indeed, be a miracle if any such thing were ever invented by a human six months old, or six years. But *adult* humans made a culture, a morality, and so on, just as naturally and inevitably as adult birds make a nest. They do not do so with the same genetically fixed specificity, that's all.

No one will believe that the normal process of human development, from conception to mature adulthood, is an artifact of education, or of culture, or of anything like that. Even William Godwin did not believe *that*, though he did believe that aging, senescence, and death are such artifacts. But it might well puzzle a more powerful mind that Godwin's to explain why the second half of human life should be determined by education or culture, while the first half is determined by the biologically given.

Nor will anyone believe that any one particular stage of normal human development is *the real person*, and that everything after that stage is just a disguise which is put on for public appearances. A normal man of 30 is not a mask which a certain child of three puts on when he thinks other people are watching; any more than a child of three is a mask which a certain infant of three months puts on when he thinks he may be under observation. The first few years of human life do, indeed, have a formative influence which is unmatched by any later period of the same length. But every stage of the entire process of development is equally real, and is not just a veneer placed over some earlier stage.

More generally, it is an advantage of the developmental perspective that it enables us to throw out that vast accumulation of stage properties and theatricality which has come down to us from the Enlightenment in general, and from the selfish theory in particular. I mean, those mountains of veils, veneers, masks, disguises, impostures, deceptions, hypocrisies, conspiracies, rackets, and the like, which Enlightened people are obliged to postulate, in order to reconcile their darling selfish theory with even the most obvious facts of human social life. You know how the stories go: religion as the creation of "monkish impostors"; morality as the self-serving invention of "artful politicians" who had got into the saddle and meant to stay there; government as deriving its existence and authority from a "social contract" which *confessedly* never existed; the medical profession as a successful conspiracy of greedy and bloodthirsty confidence men; an aversion to mother incest instilled into sons by selfish fathers for their own sexual advantage: . . . the inventory of such things is as interminable as it is incredible. Good theater can be made out of these tawdry materials, as was proved by example by Molière and Wycherley, among others, and again by Ibsen and Shaw only a hundred years ago. But they are just too silly to be even entertaining, if what you are looking for is sober truth and the biology of our species.

The Enlightenment tradition of theatricality is carried on, and of course carried further than ever before, by the sociobiologists. *Manipulation* is their favorite idea, and they find, or claim to find, manipulation going on everywhere, at every moment of human social life. No one doubts, of course, that there are such things as manipulation, hypocrisy, and self-interested lying. The only question is, how common and important they are. More specifically, the question is whether they do, or even could, permeate *all* our social life, as sociobiologists believe.

The right answer, as many philosophers have pointed out, is that they could not: that on the contrary, those things are essentially parasitic upon their opposites. Successful lying, for example, depends not only for its profitability but for its very possibility, on the existence of a general background of truth telling. But the possibility of truth telling does not depend on there being any lying at all.

It was an instance of successful manipulative communication when Brer Rabbit got himself thrown into the briar bushes he loved by telling Brer Fox that that was what he dreaded most; and no doubt similar instances are, and always have been, plentiful enough between humans. But it is not hard to see what the result would be, if in the future such manipulative communication were to become universal, or even nearly so. Communication, whether manipulative or otherwise, would then just die out altogether, for the simple reason that no hearer would ever know what any speaker meant by the words he uttered. The same obvious reasoning assures us that human communication can never have been predominantly manipulative in the past, either. A consequence is, that when the sociobiologists Dawkins and Krebs tell us in print (as quoted earlier) that all communication is "manipulation of signal-receiver by signal-sender," we would not know what they meant, if what they said were true. They *might*, after all, be secretly meaning to reduce the selfish theory to absurdity; or they might mean "A merry Christmas to all our readers"; or anything else, for that matter. But since we do in fact (to our sorrow) know what they mean, what they say is *not* true.

"All communication between humans is manipulation of signal-receiver by signal-sender"! It would not be easy to think of a viler insult to our species than this. But since the impulse to *share* our thoughts and feelings with others is *in fact* the very strongest passion of our nature, it would be hard to think of a more ridiculous one, either.

Mature adulthood, then, is not a disguise which our infantile selfishness later wears, any more than it is a veneer concealing our "savage" or our "animal" nature. We no longer need *any* of that moth-eaten stage costumery! Adults are not *hiding* their infantile selfishness: they have grown out of it, that's all. They may, indeed, be carried back to it, by torture or starvation. But so they may also be by, for example, brain damage suffered in a car accident. Yet no sane person would say, concerning a normal adult who has been "infantilized" by a car accident, that we now see him as he *really* was just before the accident happened.

AMONG HUMANS, then, as among kookaburras, rats, and indeed all the higher animals, the infants are more selfish than the adults. They are also, of course, more helpless. Nor is this conjunction of attributes accidental: there is an obvious connection between helplessness and selfishness. An animal that is helpless, whether through infancy, injury, illness, or age, if it is to survive at all, can do so only by accepting from others good offices which it cannot reciprocate: that is, only by adopting the selfish policy of "take rather than give." And the greater the degree of its helplessness, the greater must be the excess of "take" over "give" in its policy; that is, the greater the degree of its selfishness must be.

In our species, the helplessness of infants is both more extreme and more prolonged than in any other species. This fact was noticed by Anaximander about 2,600 years ago, and it suggested to him that our species must have evolved from some other: from some species which was a good deal more businesslike than ours is, in what is called in commerce "the replacement of existing stock." His observation is, indeed a most pregnant one, and its implications have perhaps not been entirely exhausted even yet. It implies, for example (since the more helpless an organism is, the more selfish it must be to survive), that human infants are also more selfish than the infants of any other species. I have never heard of any observations which contradict this corollary, or even appear to do so.

Anaximander's law (as we may call it), though it goes deep, is only a comparative proposition. It does nothing in itself, therefore, to prepare us for the *absolute degree* of infant helplessness which we find by experience in the human case. This is something absolutely staggering, indeed scarcely credible. It would be thought altogether incompatible with our species' surviving, and would be rationally thought so, if we did not, from other sources, happen to know better. Newborn humans are far more helpless, even, than (for example) the half-inch blobs which are newborn kangaroos.[31] Even after ten weeks, a baby still cannot even use its hands to guide its mother's nipple towards its mouth. As a way for the most intelligent, inventive, and capable beings on earth—and perhaps anywhere—to begin, this seems more than a little odd. Yet it *is* the way they begin, and the only way they can begin. The *bypassing* of infancy is not contemplated even by the most wildly speculative of genetic engineers.

That the infants of our species are more selfish than those of any other would be a telling blow in favor of the selfish theory, if it were taken on its own. But it has to be taken in conjunction with another fact which we know independently: that our species has survived for a very long time. And then it is a telling blow *against* the selfish theory. That our infants of each new generation are uniquely helpless and selfish, while the species has survived so long, can mean only one thing: that the adults of our species are more unselfish and helpful towards infants than are the adults of any other species.

We knew this before, of course, from everyday experience, and did not have to wait to learn it as a corollary of Anaximander's law. Nor does this corollary by any means do full justice to parental care in humans. For it is only a comparative proposition, like Anaximander's law itself. It therefore does not itself prepare us for the *absolute degree* of helpfulness and unselfishness which we find by experience that human parents bestow on their young. This is something which, like the helplessness of our young, far exceeds what could have been rationally anticipated just from a knowledge of other animals. It also far exceeds, in countless instances, any praise that words could possibly express. If there is anything about our species which could justify its existence in the eyes of a superior extragalactic spectator, it would be the amazing spectacle of our parental care. More specifically, it would be the spectacle of our *maternal* care. But let us draw a veil over a subject so painfully unfashionable . . .

Parental unselfishness and helpfulness, though the most obvious as well as the most extreme form of human altruism, is still only one form among a number of others. In recent decades, neo-Darwinian selfish theorists have attempted to find a kind of selfishness even in the parental form of altruism. The attempt does not succeed, as we will see in Essay VII below. But suppose it did. How much would that matter?

Far less than neo-Darwinians suppose. For its success would still leave untouched all those forms of altruism in which the beneficiaries are not offspring, or relatives at all, of the benefactors. These forms include the professional functions of those groups—soldiers, doctors, and priests—which are such prominent and enduring features of the human landscape, and of which I spoke at length earlier in this essay. *Their* altruism is exercised quite independently of any kinship relation between benefactor and beneficiary. Above all, I ask the reader to bear in mind the maternity ward sisters! Or, if a *maximally* unbiological example of human altruism is wanted, I remind the reader that some of history's most signal instances of military bravery and *esprit de corps* were furnished by units of the Spartan army which were exclusively homosexual.

Even after all of that, there remains the colossal fund of non-biological altruism which, in advanced societies of the present day, is crystallized in the taxation system, and the ends which it exists to serve. Could *this* be selfishness, or selfishness qualified only by the parental altruism which we share, to a greater or less degree, with the adults of all the higher animals?

No. To anyone not utterly blinded by a theory, it is perfectly obvious that on the contrary our species, even apart from kinship, is sharply distinguished from all other animals by being in fact *hopelessly addicted to* altruism. It will be time to think otherwise when, and not before, adult wolves or kookaburras or rats pool their resources in order to relieve the illness, or improvidence, or ignorance, of conspecifics to whom they are unrelated. And *that* (as old Australians say) will be the day.

Essay 7
Genetic Calvinism, or
Demons and Dawkins

. . . these puppets [that is, people and all other organisms] are not
pulled from outside, but . . . each of them bears in itself *the clockwork*
from which its movements result.
—Schopenhauer, *The World as Will and Representation*

I

SUPPOSE THAT J. S. Bach had been very rich when he died, and
had provided in his will for a valuable scholarship to be
awarded each year to the most gifted young composer that could
be found. Or suppose Isaac Newton had been rich all his life, and
had at one time or another supported at his own expense various
talented but poor young mathematicians. Would these have been
selfish actions on the part of Bach or Newton?

Clearly not: quite the reverse, in fact. Their actions would have
been thought, and rightly thought, to be decidedly *un*selfish ones.
They would have been praised, and rightly, as evidence of Bach's
devotion to music, or of Newton's disinterested love of knowledge. There might be evidence that Bach and Newton were selfish
men, but *these* actions could not possibly form part of that
evidence, since they are plainly evidence to the contrary. Human
nature being what it is, there might indeed have been some tinc-

ture of *vanity*, in their performing generous actions on the conspicuous scale of these endowments. But vanity is not at all the same thing as selfishness. It is not even unusual for an unselfish person to be also a vain one.

Nowadays, however, there are certain neo-Darwinian biologists who would say that these actions *were* selfish ones, because of their "self-replicatory" tendency. That is, because Bach and Newton, by doing these things, had adopted the best means open to them, with the possible exception of parenthood, of increasing the number of people *like themselves*. What should we think of someone who said this?

Badly, anyway. First, he would be deliberately making moral mischief. For "selfish" is a term of opprobrium, and anyone who applied it to these actions of Bach or Newton *must* tend to make people think the worse of those men, on account of certain actions which were in fact greatly to their credit. And anyone who knows enough to be a biologist is sure to *know* that "selfish" is a term of opprobrium; so that this biologist would know he was making this moral mischief.

Second, he would be making intellectual mischief. For nothing whatever can literally replicate *itself*. The most that anything could possibly do in that way would be to produce perfect copies of itself. By contrast, the object or target of selfishness is—by the very meaning of that word—*oneself*, and nothing else. Superscientist may create in his laboratory an exact replica of me, or I may happen to have an identical twin. But it is not this copy or twin who is the object of *my* selfishness: it is myself. This copy or twin will plainly be nothing at all to me if, as could happen easily enough, I do not know of his existence. If I do know of him, he may be much to me, or little, or again, nothing at all. But one thing that he cannot possibly be is the object of my selfishness: namely me.

In reality, of course, the tendency of Bach scholarships to produce Bach replicas would be extremely weak. But let it be supposed to be as strong as you like: suppose that, in some mysterious way, a Bach scholarship always transformed the recipient of it into an exact replica, mental as well as physical, of J. S. Bach at the age of twenty-two. Would this mean, or would it be even the slightest evidence, that in creating his scholarships Bach had behaved selfishly? Again, obviously not.

It is certainly some evidence of vanity, if a man multiplies copies of a picture or a statue of himself. But vanity (as I have said) is not selfishness, and multiplying copies of oneself is a very different thing from multiplying pictures or statues of oneself. If a man happens to have ten sons who are all extremely like him, that is not the slightest reason to believe that he is selfish. If anything, it is some faint evidence that he is not. At any rate, it is well known that selfishness is something which often deters people from having any children at all.

Viruses have a strong tendency to self-replication: but what would we think of a virologist who, on that account, insisted on calling viruses "selfish"? Well, this virologist, unlike someone who said that Bach's creation of his scholarships was selfish, would not be making moral mischief. But he would be deliberately making intellectual mischief, in two ways. First, by applying a term of opprobrium to behavior which, since it is the behavior of viruses, cannot intelligibly be made the subject or either opprobrium or praise. Second, by saying something which, even apart from all questions of praise or blame, does not make sense, and which he knows does not make sense.

Viruses not only *are* not selfish: they could not be. It makes no sense to say of a virus that it is selfish, any more than to say of a virus that it is (for example) studious, or shy. You could just as intelligibly describe an electron as being slatternly, a triangle as being scholarly, or a number as being sex mad. And this is a fact which could not fail to be known to anyone educated enough to know what the words "virus" and "selfish" respectively mean: a condition which is sure to be satisfied by anyone educated enough to be a virologist. So any virologist who insisted on calling viruses "selfish" would be insisting on saying something which he himself knows does not make sense. And if this is not deliberately making intellectual mischief, it will do as an example until the real thing comes along.

Genes, like viruses, have a strong tendency to self-replication. But to describe genes as "selfish" on that account, or on any account, would be just as nonsensical as describing viruses as "selfish." Genes can no more be selfish than they can be (say) supercilious, or stupid. Yet while no real life virologist ever *has* called viruses "selfish" (as far as I know), there really is a geneticist who does insist on describing genes as "selfish."

This is Dr. Richard Dawkins, of Oxford University, and to say that he insists on talking in this way is to understate the case extremely. He wrote a book which purports to explain evolution as principally due to what he calls the "ruthless selfishness" of genes. And, as if in order to exclude all charitable misunderstandings, he actually entitled his book *The Selfish Gene*.[1]

Since it is not only nonsense, but very obviously nonsense, to say that genes are selfish, it might reasonably have been anticipated that the publication of this book would injure Dr. Dawkins's scientific reputation. But in fact the effect was the very reverse. *The Selfish Gene* not only became a best seller, but at once elevated its author into the very front rank of biological authorities: a position which he enjoys to this day.

Surely there is something in this more than a little puzzling? Imagine, reader, that you or I wrote a book with a transparently nonsensical central thesis which was crystallized in its title: say, *The Sex Mad Prime Numbers*. How far, do you think, would our proposed book get with the readers which publishers employ? Even if it were published, how far do you think it would get with the public? Not very far, anyway. And yet *The Selfish Gene* was an immense success not only with lay but with learned readers. How is this prodigy in literary history to be explained? I believe I can answer this question.

ONE OF THE PIONEERS of genetics, William Bateson, was fond of repeating a remark which a Scotch soldier made to him during the 1914–18 war, after listening to one of his lectures: that genetics is "scientific Calvinism."[2] Well, what Dawkins did in *The Selfish Gene* was in effect to embrace this old joke, or three-quarters joke, as being no joke at all, but the sober truth. Genes are to him what demons were to Calvinist theologians in the sixteenth century, or what "Zurich gnomes" used to be to socialist demonologists of our own century. That is, they are beings which are hidden, immoral, and invested with immense power over us: power so great, indeed, that we are merely their helpless puppets, except insofar as God, or History, or some equally extraordinary causal agent comes in to assist us.

Calvinist theology, in its strict form, denies that any created thing has any causal powers at all. God is supposed to be the one and only cause of anything and everything in the universe. All created things are mere epiphenomena: effects, not causes. But, as might have been foreseen, Calvinists were never able to adhere to this position consistently. The reason was that they considered themselves charged with a most momentous mission: to enlist their fellow men on God's side, in the cosmic war against Satan and all the other fallen angels. That is, against devils.

Calvinists were therefore obliged, from the very start, to admit that there *is* one class of created things which does possess causal powers, and appallingly great causal powers at that: namely, devils. This, however, was plainly inconsistent with the strict letter of their theory. They therefore had to go in for a great deal of unsightly squirming. The squirmings of Calvin himself, for example, are positively painful to watch. He tells us that devils are an inexpressible danger to every human soul, but also tells us that no devil can ever win. He says that devils are God's enemies, and are most potent causes of evil, but also says that they can do nothing except by God's permission and appointment.[3] And so on.

DAWKINS IN *The Selfish Gene* is not, of course, engaged on any mission of cosmic warfare or of moral reformation. But just as Calvin divides created things into potent demons and causally impotent everything else, so Dawkins divides the organic world into potent genes and causally impotent everything else. According to Calvinism, *we are pawns in a game*, in which the only real players are the demons and God. According to *The Selfish Gene*, we are pawns in a game in which the only real players are genes.

You, your dog, and the plants in your garden are causally null, according to Dawkins, or at any rate negligible: only some throwaway envelopes, of a fleshy or fibrous composition, which it suits certain genes to make brief use of, as they go irresistibly about their everlasting business of making still more copies of themselves. Organisms are merely "fronts" for the genes which sit inconspicuously inside them, just as Capone staff used to sit inconspicuously in Chicago betting shops and "dance parlors," or as well disguised Politburo staff used to sit in the ruling bodies of Western "peace" movements.

The branches of literature are very various, and the readers in one branch tend to be not readers in the others: the readers of science, say, tend not to be readers of history, or of philosophy, or of poetry. But there is a branch of literature which, at one level or another, finds favor with *all* readers. This is, books of revelations of "wickedness in high places," as St. Paul says: books which disclose the appalling immorality which is rife among those who are placed furthest above us in power, and whose activities are, as a general rule, the most completely hidden from our view. "The Secret History of the Court of King So-and-So" has been a natural born best seller ever since the time of Procopius, 1,500 years ago, and no doubt much longer still. Such histories are best sellers at this very moment, concerning the British royal family. Books like these *cannot* fail. For they give the reader "the life styles of the rich and famous," and lavish helpings of real-life violence, sex, manipulation, selfishness, and greed. Who could ask for anything more?

Now, genes are *in fact* extremely well hidden: so well hidden, indeed, that before the present century not a single human being ever knew of their existence. Then, genes *are*—it has turned out—causal agents of immense power, in human as in animal and plant life. Finally, the immorality of genes is extreme; at least, it is, if the report which Dawkins gives of them is true. For he describes them as being ruthlessly selfish. And he believes, as most of us do, that ruthless selfishness is extremely immoral, and something which it is imperative to discourage in ourselves, in our children, and in others.

This is the explanation of the runaway success of *The Selfish Gene*. The book did not contain any addition to existing knowledge or theory in evolutionary biology. Indeed (except for its last chapter, of which I shall speak later), it did not even claim to do so. It was *avowedly* a book which expounded, combined, and semi-popularized the main contributions which others had made to evolutionary biology in (roughly) the preceding forty years: say, since R. A. Fisher's *The Genetical Theory of Natural Selection* (1930). But Dawkins had the wit to perceive, as no one had before him, that genes, since they are hidden, powerful, and immoral, furnished the materials for a book of "Secrets and Scandals of the Court of King Gene." No power on earth could have prevented such a book from succeeding.

DAWKINS MORE THAN ONCE assures his readers that when he says genes are selfish, he is not nonsensically attributing to them a certain psychological or "subjective" character. He does not mean, he says, that genes are "conscious, purposeful agents."[4] Applied to genes, the language of selfishness is "only a figure of speech."[5] But he finds it a help in conveying to his readers, what he believes to be literally true, that organisms are simply certain vehicles which genes design, build, and manipulate, as part of the longer term process of increasing the number of their own copies. Anyway, he says, calling genes "selfish" cannot be importantly wrong, because it is *dispensable*. We could always "translate [it] back into respectable terms if we wanted to."[6]

The sense in which he uses the word "selfish," Dawkins writes, is one which is standard in biology, and which is "*behavioral*, not subjective."[7] It is this. "An entity, such as a baboon, is said to be altruistic if it behaves in such a way as to increase another entity's welfare at the expense of its own. Selfish behavior has exactly the opposite effect. 'Welfare' is defined as 'chances of survival' . . . ."[8]

It is true that this is the standard sense in which neo-Darwinian biologists use the words "selfish" and "altruistic" respectively. It is also true (as we saw in Essay 6) that it is a problem or worse for neo-Darwinism (as for Darwinism) how altruistic behavior could survive and spread in any population of animals. But let all organisms be as selfish as the extremest neo-Darwinian cares to suppose: that would still not justify anyone in calling *genes*, as distinct from organisms, selfish.

Yet Dawkins says he uses the word "selfish" in the behavioral sense (as we have just seen), and he *will* have it that genes are selfish. But what connection is there, between selfishness in the behavioral sense, and that feature of genes on which everything in *The Selfish Gene* turns: their self-replicatory propensity? To justify his calling genes selfish in the behavioral sense, Dawkins would need to show that self-replication increases the self-replicator's chances of survival. But how on earth could he, or anyone, possibly make *that* out?

My identical twin, or a laboratory-made replica of myself (as I pointed out earlier), is not a possible object of my selfishness, in the ordinary psychological sense of "selfishness." But suppose that I am myself Superscientist, and that I manufacture my own replica or twin. Have I then done something selfish, even in the behavioral sense of "selfish"? Have I improved my own chances of survival at the expense of the chances of others?

It is perfectly obvious that I have not. The coming into existence of a perfect copy of myself might, just conceivably, tickle my vanity. But it would not remove one year or one second from my age, or lighten, by however little, the burden of my present or future illnesses or other afflictions. My age, health, wealth, and prospects would be just what they were before I conjured up my replica. Any rational insurance company, and any rational person, would tell you the same thing. And since I have *not* increased my own chances of survival, I have certainly not done so at the expense of anyone else's chances.

Equally plainly, the same is true of genes. By making a copy of itself, a gene certainly does not gratify its selfishness in the ordinary sense of that word, since (as I said earlier) genes cannot *be* selfish in that sense. But neither does it do anything selfish in the behavioral sense. Self-replication would even seem (to a layman such as myself) rather to *worsen* a gene's chances of survival, since it must use up a sizable part of its limited energy store. But even if that is merely a layman's misunderstanding, it seems obvious enough that a gene, by self-replicating, does not *improve* its own chances of survival. (Its replica is not going to look after the parent gene in its old age, for example.) Which is to say, that the self-replication of a gene is *not* selfish, even in the sense in which Dawkins says he is using that word.

At this point, however, Dawkins would remind me that "the selfish gene . . . is not just one physical bit of DNA . . . it is *all replicas* of a particular bit of DNA, distributed throughout the world."[9] What a gene does by self-replicating, he says, is to benefit "itself in the form of *copies* of itself."[10] "The gene is a long-lived replicator, existing in the form of many duplicate copies" of itself.[11]

There: you have just witnessed how Dawkins made out the case on which his whole book depends. How he managed, that is, to represent the self-replication of genes as being selfish in the behavioral sense. Well, there is nothing to it, really, once you have seen how the thing is worked. All you need to do is to talk about things which *exist in the form of other things,* and more specifically about things which *exist in the form of copies of themselves*; and the job is done.

Talking like this may seem at first sight to be only an innocent departure, indeed only a trivial departure, from ordinary ways of speaking and thinking. But a little further reflection will soon correct that initial impression. The truth is that Dawkins has here done much more than sum up recent progress in evolutionary biology. In fact, he has opened up unlimited vistas of future intellectual and even economic progress, in very many fields.

For example, Dr. Dawkins should certainly say to his identical twin (if he has one): "In your own interests you ought to give me all your money, because by doing so you would benefit yourself in the form of a copy of yourself." His brother will selfishly embrace this novel way of enriching himself, if the biology of *The Selfish Gene* is true; while at the same time the advantage which will accrue to Richard Dawkins is also clear. As a solution to a problem which must often arise between identical twin brothers this must be admitted to be as ingenious as it is equitable.

Then, think of the doctrine of the Trinity in Christian theology, and of the agonizing perplexity which it has caused to thoughtful Christians for two thousand years. All of that perplexity can now be made a thing of the past, with just one touch of the logic of gene selfishness. There is God Himself, the Father. But He also exists in the form of two copies of Himself: the Son, and the Holy Ghost. What could be more simple than that, or more satisfactory? If existing in the form of copies of oneself is so easy and uncontroversial that mere genes accomplish it all the time, it cannot possibly be too hard a task for members of the Trinity. We can therefore look forward, as a result of *The Selfish Gene*, to the early extinction of all Trinitarian controversy, and in particular to a rapid healing of the tragic schism which has divided Western and Eastern Christianity for almost a thousand years.

Again, there are many people, beyond doubt, who would pay good money to have Elvis Presley in the house. There is a fortune to be made by the first manufacturer who benefits from reading *The Selfish Gene*, and mass produces Elvis Presley in the form of Presley dolls of some acceptably high degree of fidelity to the original. Or, to be strictly accurate (and in order not to encourage groundless hopes of easy wealth): there is a fortune to be made in Presley dolls *if* it is true that something can exist in the form of copies of itself, and *if* it is true that a gene's self-replication is selfish in the sense of increasing its own chances of survival.

IF YOU CANNOT, without fudging, get from self-replication to self ishness even in the behavioral sense, then you certainly cannot get from self-replication to selfishness in the ordinary psychological sense. And yet it is not really open to doubt that it was the ordinary sense of the word which, though repeatedly disavowed by the author, really "carried" Dawkins's book with his readers. Suppose that, before publishing it, Dawkins actually had done what he says is always open to him, or to anyone to do: translate every reference in the book to selfishness "back into respectable terms," about self-replication. What would have been the result? The title of the book would have become *The Self-Replicating Gene*: which is about as interesting as watching paint dry, or as entitling a book about cats, *The Fish-Eating Cat*. And in the process of translation, every suggestion of revelations being made about wicked, powerful, and hidden rulers, would have been lost. Without these allurements, the book would have fallen "dead-born from the press." Or at best, it would have made no greater public impact than (for example) G. C. Williams's *Adaptation and Natural Selection* (1966): a book of very similar scientific content to *The Selfish Gene*, but of far greater merit, which was never anything remotely like a best seller.

Nor did the author of *The Selfish Gene* differ, in this respect, from his readers. For him, as for them, it is the ordinary psychological sense of "selfish" which gives his book its interest. If this were not so, it would be quite impossible to explain, for example, the paragraph on the upper half of page 2 of *The Selfish Gene*. For here, even though he has not yet even introduced the distinction between the psychological and the behavioral senses of the word, Dawkins calls both genes and human beings "selfish" in the same breath. Would he have done that, do you think, if he had been anxious to avoid being misunderstood as saying that genes *are* selfish in the ordinary sense?

If the question were asked, then, whether Dawkins really believes that genes are selfish in the ordinary sense, the answer best supported by the text of his book would be: "of course he doesn't; yes he does." This inconsistency was complained of by a philosopher, Dr. Mary Midgley, in the course of a scathing attack on *The Selfish Gene*. She said that Dawkins seemed to have acquired "the useful art of open, manly self-contradiction."[12] But a better explanation of Dawkins's inconsistency, and one which is a fraction more sympathetic, is not far to seek.

II

I do not believe that humans are the helpless puppets of their genes, and cannot even take that proposition seriously. Why? Because I have heard far too many stories like that one before, and because it is obvious what is wrong with all of them.

"Our *stars* rule us," says the astrologer. "Man is *what he eats*," said Feuerbach. "We are what our infantile sexual experiences made us," says the Freudian. "The individual counts for nothing, his class situation for everything," says the Marxist. "We are what our socioeconomic circumstances make us," says the social worker. "We are what Almighty God created us," says the Christian theologian. There is simply no end of this kind of stuff.

What is wrong with all such theories is this: That they deny, at least by implication, that human intentions, decisions, and efforts are among the causal agencies which are at work in the world.

This denial is so obviously false that no rational person, who paused to consider it coolly and in itself, would ever entertain it for one minute. No one ever doubts, at least while he has or remembers having a big fish on his line, that the intentions and efforts of even a fish can make a difference to the outcome of a situation; especially if the fish gets away after all. And if even fish efforts sometimes have causal efficacy, then human efforts can hardly be altogether without it.

The falsity of all these theories of human helplessness is so very obvious, in fact, that the puppetry theorists themselves cannot help admitting it, and thus are never able to adhere consistently to their puppetry theories. Feuerbach, though he said that man *is* what he eats, was also obliged to admit that meals do not eat meals. The Calvinistic theologian, after saying that the omnipotent Creator is everything and his creatures nothing, will often then go on to reproach himself and other creatures with *disobeying* this Creator. The Freudian therapist believes in the overpowering influence of infantile sexual experiences, but he makes an excellent living by encouraging his patients to believe that, with his help, this overpowering influence can be itself overpowered. And so on.

In this inevitable and tiresomely familiar way, Dawkins contradicts *his* puppetry theory. Thus, for example, writing in the full flood of conviction of human helplessness, he says that "we are . . . robot-vehicles blindly programmed to preserve the selfish molecules known as genes,"[13] etc., etc. But at the same time, of course, he knows as well as the rest of us do, that there are often other causes at work, in us or around us, which are perfectly capable of counteracting genetic influences. In fact, he sometimes says so himself, and he even says that "we have the power to *defy* the selfish genes of our birth."[14] As you see, he is just like those writers of serial stories in boys' magazines, who used to say, in order to extricate their hero from some impossible situation, "With one bound, Jack was free!" Well, it just goes to show that even the most rigid theologian of the Calvinist-Augustinian school has got to have a Pelagian blow-out *occasionally*, and deviate towards commonsense for a while.

Here is another specimen of Dawkins contradicting his own theory. He says, "let us try to *teach* generosity and altruism,"[15] but also says that "altruism [is] something that has no place in nature, something that never existed before in the whole history of the world."[16] Well, I wonder where we are, if not "in nature"? And (as Midgley pertinently asked), who are Dawkins's "us": the ones that are to teach altruism? Principally parents, no doubt. Well, parents are not, what Dawkins implies they are, just some shoddy temporary dwellings rigged up by genes. But neither are they creatures from beyond, "sidereal messengers," or sons and daughters of God sent down on a mission of redemption and reformation. Parents are just some more people and hence, if you believe Dawkins, are selfish. Where are they, on his theory, to *get* any of the altruism which he wants then to impart to their children? And as for altruism having "never existed before": one longs to learn, before *when*? Before *Homo sapiens*? Before the eighteenth-century Enlightenment? Before the British Labour Government of 1945? Dawkins should not have omitted to tell us at least the approximate date of an event so interesting, and (apparently) so recent, as the nativity of altruism.

ALTHOUGH WE human beings are fully paid up causal agents, we enter into very *unequal* causal partnerships. Sometimes we are the senior partner in these, sometimes the junior. A man is certainly the senior partner of his puppet—I mean a real wood and string puppet. Likewise in the causal partnership between a man and a car; if both are in normal working order, the man is the boss. But we are also often junior causal partners, indeed very junior ones, even to other human beings. In business, or politics, or personal affairs, another man may even be able to prevail upon me to act simply as his "pawn" or "puppet."

But even the most junior of causal partners always has some powers of its own, and some powers, at that, which it exercises upon the senior partner. Even the lowliest of human pawns must possess the powers of speech, of movement, and so forth; and he would not have got the pawn job in the first place, if these powers had not made at least some impression on the man who makes use of him. Your car is the *locus* of countless causal powers which are independent of your will, and every now and then it reminds you unpleasantly of this fact. Nor could you drive the car at all, unless you constantly received from it "signals" which inform you of its current state. A wooden puppet is one of our most junior causal partners, but even it cannot be got to do what you want it to, unless you first ascertain its *present* state; which means, unless its present state affects your senses. Even then, manipulating a puppet depends upon its retaining its own causal bent, including its propensity to gravitate. If you were in the middle of a long space journey, and in a state of weightlessness, you could not help to pass the weary hours by putting on a puppet show.

FOR REASONS like the obvious ones which I have now given, sensible people take no notice, when yet another crank or charlatan publishes yet another book which says that human beings are the helpless puppets of something or other: God, or God and demons, or History, or Race, or the Unconscious, or Aliens from Outer Space, or whatever. *The Selfish Gene* is simply another member of this slum breed of books, and ought to have been recognized as such from the start.

But in fairness to Dawkins, I need to add that *genetic* puppetry is the most excusable of the bad breed of theories to which it belongs. It is certainly better than Freudian puppetry theory, for example, or Marxist puppetry theory. In fact there are two partial excuses for it, neither of which is available to *any* other puppetry theory.

One is that although genetics is not yet even one hundred years old, it has already revealed the existence of a previously unknown class of causal agents, which are so powerful that the discovery of them has left us in a state of shock and fear, mingled with intellectual intoxication. Genes—of which Charles Darwin died as ignorant as Julius Caesar did—have turned out to be, in many respects, very senior causal partners indeed in the making of ourselves and all other organisms. We now *know* of many human attributes whose causation is entirely or principally genetic. In these circumstances, is it any wonder that some people have made a demonology out of genetics? On the contrary, it would have been a wonder is no one had done so.

The second excuse or extenuating circumstance for genetic puppetry theory is this: that there has been a great and effective conspiracy, during most of this century, to prevent the knowledge of genetics from being publicly diffused. The two principal conspirators were originally, of course, Joseph Stalin and T. D. Lysenko. Although they are long dead, their conspiracy lives on, its main centers now being the humanistic departments of Western universities. The very first priority in these departments is to conceal from their students the importance, or even the existence, of genetics, especially human genetics. This conspiracy has actually reached quite new heights in the last five years, with the triumph in universities of "political correctness." Nowadays, in universities, human geneticists had better keep their mouths shut;

and so, of course, they do. That is why they are moving, in droves, into non-university research institutions.

The Selfish Gene is very nearly as bad a book as Mary Midgley said it is, and it is quite as pernicious a book as she said it is. It is so, moreover, very largely for the reasons she gave in her admirable article about it. But that article gives no indication, as it should have done, that genetic puppetry theory is any better than the puppetry theory of (say) a sixteenth-century witch finder. Well, it is not saying much, but it is certainly better than that.

III

It is no mystery why the supply of puppetry theories never fails: there is an unfailing *demand* for them. People want relief from responsibility, and puppetry theories promise them this relief. There is also (as Mary Midgley pointed out)[17] a sadomasochistic element in all such theories. People take a certain *satisfaction* in contemplating their own supposed helplessness, and the irresistible power of their masters. Puppetry theories appeal to some degree to everyone, because everyone has, to some degree, this yearning for irresponsibility and a taste for sadomasochism. But those two things must clearly be especially strong in someone who *invents* a puppetry theory: you need an unusually "demonological" cast of mind to do that.

For this reason puppetry theories, in the hands of the few individuals who invent them, always display a strong tendency to *expand*. The man who has dreamed up a set of demons or puppet masters behind one field of phenomena is quite the likeliest man to dream up, later on, *another* set of demons behind another field of phenomena; or to come up with a single, but far wider set of demons, comprehending the set which he had happened to stumble upon first. The people who suffer from delusions of being conspired against are always being obliged to conclude that this conspiracy is more widespread than they had previously realized. In the sixteenth and seventeenth centuries, witch finders were constantly and genuinely astonished, because where they had expected to find one witch, they *always* found twenty.

It is therefore not surprising, though it is certainly alarming, when Dawkins announces, in the last chapter of *The Selfish Gene*, the existence of a *second* set of hidden beings who manipulate us while selfishly replicating themselves. This was a scientific bombshell; or anyway a bombshell. It is the only original part of the book, which up to this point had consisted of exposition, spiced only with sensationalism about selfishness, or other people's discoveries or theories. But whereas genes had been known for most of the century, the existence of this second set of demons was entirely unknown until Dawkins revealed it in his last chapter.

That chapter is the worst part of the book, by a margin which it would be difficult to exaggerate. But so extremely favorable has been the book's overall reception that even this part of it has met with respectful attention from many grave professors. Their most frequent reaction to it, all too predictably, has been of the "It's an outrageous proposal but we'll certainly consider it" kind.[18] But as recently as 1989, at least one very distinguished American philosopher has given this part of the book his enthusiastic assent.[19]

When genes were first discovered, they had, of course, no established name. The discoverers' phraseology varied, and an entirely new word was obviously needed. A Danish friend of William Bateson suggested "genes," Bateson himself coined "genetics," and these names stuck. Just so, Dawkins needed a new name for the new things which *he* had discovered in 1976. He decided to call them "memes."

A meme is anything which can be transmitted by non-genetic means from one human being to another. Hence all ideas, beliefs, attitudes, styles, customs, fashions—in fact all the elements of culture in the broadest sense—are memes. There is a meme for Pythagoras's Theorem, and another for wearing stiletto heels; a meme for being in favor of capital punishment, and one for the idea of a triangle; a meme for the Mozart Requiem and another for shaving...

Now, Dawkins says, organic evolution is driven by the struggle between one gene and its rival genes for a place on the chromosome, and with that, the chance to self-replicate; and just so, *cultural* evolution, he says, is driven by the struggle between one meme and its rival memes for a place in our *brains*. Take, for example, the meme for the belief that the sun is at the center of the local planetary system. A few brains in classical antiquity had contained this meme, but it then disappeared for nearly two thousand years. In the mid-sixteenth century, however, it popped up again in the brain of Copernicus, and a struggle began between this heliocentrism meme and the geocentrism meme. At that time, the latter was settled in almost all brains, but the heliocentrism meme has won this struggle long ago. It has been so successful, in replicating itself from one brain to another, that by now there are hardly any brains left which contain the geocentrism meme.

Even a single kind of meme, such as a belief meme, can be transmitted from brain to brain in many different ways. Beliefs can be transmitted, for example, by teaching people some science or mathematics—heliocentrism or Pythagoras's Theorem, say; by telling people lies; or by brainwashing them. Now, we usually think of these as being three importantly different activities; and we think of them as all being, in any case, activities of human *agents*. But again, according to Dawkins, all of that is a mistake. Teaching science, lying, and brainwashing are simply three different ways in which memes in certain brains succeed in replicating themselves in other brains. And in all three alike, the causal agents at work are not human beings: it is the memes *themselves* that do these various things.

Well, that is what Dawkins discovered in 1976, if "discovered" is the right word. His "discovery" of memes comes as no surprise to anyone who has read all the preceding chapters, on genes, and has noticed the strongly demonological cast of Dawkins's mind. Puppetry theories, as I have said, *always* tend to expand, and Dawkins was therefore always likely to reenact his "discovery" of selfish genes, or to do it again on a bigger scale. Ideally, no doubt, he would have preferred to have just one *super* giant-sized conspiracy, to explain at once biology *and* culture. But not seeing his way to that, he insisted on at least having two giant-sized conspiracies, one for biology and one for culture. Genes are not ruthlessly selfish, but Dawkins is certainly ruthlessly demon-ological. You could put him down anywhere in the world and rely on him to find there, what no one had before, invisible pup-pet masters manipulating visible puppets. If, in addition, these puppet masters should possess any natural tendency towards self-replication, he would be sure to repeat his absurd though profitable trick of calling them "selfish" on that account. If he ever turns his mind to cosmology or fundamental physics, we can be confident of his making "discoveries" there which are even more valuable than those of memes and selfish genes, though (alas) of the same general kind as those two.

Are memes a scientific discovery? Well, one thing is absolutely certain; if they are, they are the most *effortless* scientific discovery of all time. For what did it take, after all? What was the evidence and the reasoning that enabled Dawkins to discover memes in 1976, although their very existence, like that of genes before 1900, had been unsuspected before?

Well, to tell the truth, it was nothing more than the following:

Sometimes such things as beliefs, attitudes, etc., are transmitted non-genetically from one person to another.

So,

There are memes.

I can only echo Huxley's famous remark after he first read *The Origin of Species*: "How extremely stupid not to have thought of that!" Why, even *I* had known for years before 1976 that people often "pick up" opinions, attitudes, etc., from other people to whom they are not related. So, if memes are indeed a scientific discovery, I must myself have stood on the brink of a place in the history of science! Only, alas, I did not know it, and in the finish,

Dawkins got the "glittering prize" before I or anyone else realized that it was something valuable.

There was, of course, this much excuse for me and all the other losers; that Dawkins's discovery of memes is utterly unlike anything which the history of *science* has made us familiar with. What scientific discovery ever looked like *that*? And yet there is something familiar about Dawkins's discovery, at any rate to a philosopher: something horribly familiar, in fact. I have seen that kind of thing hundreds of times before, but where? Why, in these absolutely effortless pseudo-discoveries that philosophers make, and on which their fame rests. Plato's "discovery" of universals, for example, or Kant's "discovery" that existence is not a property.

Plato's discovery went as follows.

> It is possible for something to be a certain way and for something else to be the same way.

So,

> There are universals.

(*Tumultuous applause, which lasts, despite occasional subsidences, 2,400 years*)

Kant's "discovery" went thus.

> Any property that a real x had, an imaginary x could have, and any property that an imaginary x could have, a real x could have.

So,

> Existence is not a property.

(*Hearty applause, maintained steadily for 200 years so far.*)

This kind of maximum effortlessness is typical of philosophical "discoveries," and it is not hard to say *which* kind it is: the elements of it are obvious enough. The premises must be of minimum number (ideally one) and each premise must be of maximum triviality. The line of the reasoning must be of minimum length, and be strictly deductive, or of zero risk. This last desideratum is most easily achieved, and often is achieved, by simply making the conclusion just a more arresting way of saying what the premise had said to begin with. The Plato "discovery" is a case of this. "Universals" is simply the name philosophers give to the ways in which two or more things can be the same.

Similarly "memes" is just the name that Dawkins coined for the things which humans can communicate to one another nongenetically. In fact, Dawkins's "discovery" of memes satisfies perfectly all the above requirements for a philosophical "discovery," with just one partial exception. His premise is certainly trivial enough to satisfy most people's appetite for triviality: but philosophers are rather more exacting in that matter than most people. Dawkins's premise goes so far as to assert that a certain process actually happens; whereas a typically philosophical premise, such as Plato's or Kant's, says no more than something or other is *possible*.

Yet *scientific* discoveries, it can hardly be necessary to emphasize, are exactly the opposite of philosophical pseudo-discoveries in every respect. Their premises are many, and none of them trivial: to establish even one of them can easily take years of painstaking experimental work. The path of the reasoning is not only long but complicated, and every one of its major lines possesses great internal complexity of its own. In fact, the detailed "map" of a journey of scientific discovery is always so long and complicated, that hardly anyone will ever be even able to *follow* it all *afterwards*; yet doing that is, of course, a great deal easier than *making the journey* in the first place! Finally, the reasoning which terminates in a scientific discovery remains incurably inductive after all. For some at least of the premises will be observation statements, whereas the conclusion will go further than any observation, in breadth, or depth, or both.

But the pseudo-discovery of "memes" invites a more specific

comparison, with the real scientific discovery of genes. Dawkins, to make his "discovery," did not need to draw upon any specialized knowledge, or to exercise either his experimental or his inferential powers. All he needed was to remember that some things are transmitted non-genetically from one person to another, to give these things a new name, and then to allow free rein to the demonological bias of his mind. It was absolutely effortless. In fact it is so easy to discover memes that a disciple of Dawkins, writing in a recent coffee table book, is able to make his readers perfectly *au fait* with them, in about eight lines of print and four seconds.[20]

The discovery of genes, by contrast, was remarkably long drawn out in time. It extended from Mendel's work in the 1860s on crossing various strains of peas, through the rediscovery of that work in 1900, to at least the early breeding experiments of T. H. Morgan during the First World War. Now, was any of *this* effortless? Surely, on the contrary, Morgan and his associates had first to acquire a good deal of biological information, and then work rather hard and long with their heads and hands, to design, perform, and interpret their experiments? In 1900 Bateson perceived, though few other people did, what Mendel's experiments on peas really meant: and I suppose that this difference between Bateson and most other people *must* have had something to do with his vast fund of biological information, and with prolonged and severe exercise of his penetrating intelligence.

But in all of this, easily the greatest feat of intellectual penetration was that of Mendel himself. The phenomena of inheritance are so bewilderingly various, that no one before Mendel, not even the most expert breeders of plants and animals, had ever been able to "see the wood for the trees." Yet in order to be understood, these phenomena only required to be looked at in the light of two ideas—that the "factors" of inheritance do not blend in the offspring, and that they assort themselves independently of one another—ideas which, as R. A. Fisher suggested,[21] had been as available to anyone, for thousands of years, as they were to Mendel. What a certain car rental firm claims to do, Mendel did—he *tried harder*: he concentrated his mental gaze for years on the vast jumble of apparently meaningless ratios of inherited characteristics in his peas, until he obliged these speechless witnesses to yield their secret.

During his life, of course, and for sixteen years after his death, Mendel's achievement went not only unappreciated but unnoticed. If only, now, he could have had a *Dawkins* to advise him on literary marketing! But this comparison, between the laborious but glorious scientific discovery of genes, and Dawkins's effortless philosophical pseudo-discovery of "memes," is too painful to be pursued for long. It excites too much indignation and contempt for the latter.

BUT DAWKINS'S CHAPTER on "memes" also excited in me a good deal of alarm. The demonological cast of mind runs easily (as is well known) into mental disorders of a very dreadful kind, and little amenable to treatment. Among the symptoms of these disorders there are none more common than delusions of being "possessed" by "evil spirits," or of being "occupied" by "alien forces," or of being "parasitized" by hostile organisms as yet unknown to terrestrial science. And then, I read the following expression of the meme theory, written by a colleague of Dr. Dawkins, but heartily endorsed by him.

Memes are "living structures, not just metaphorically but technically. When you plant a fertile meme in my mind, you literally parasitize my brain, turning it into a vehicle for the meme's propagation in just the way that a virus may parasitize the genetic mechanism of a host cell. And this isn't just a way of talking—the meme for, say [Pythagoras's Theorem] is actually realized physically, millions of times over, as a structure in the nervous systems of individual men. . . ."[22]

I cannot speak for others, but for my own part, it is impossible to read these words without feeling anxiety for Dr. Dawkins's sanity. I try to think of what I, or anyone, could say to him, to help restrain him from going over the edge into absolute madness. But if a man believes that, when he was first taught

Pythagoras's Theorem at school, his brain was parasitized by a certain micro-maggot which, 2,600 years earlier, had parasitized the brain of Pythagoras, . . . what *can* one say to him, with any hope of effect? And if a man already believes that genes are selfish, why indeed should he not also believe that prime numbers are sex mad, or that geometrical theorems are brain parasites?

One might try saying to Dr. Dawkins: "Look, you are in the phone book, and they print millions of copies of the phone book—right? But now you *don't* believe, do you, that you are there millions of times over 'in the form of' printed letters, or 'realized in' the chemistry of ink and newsprint?" But I would be so afraid of being told by Dr. Dawkins that he does believe this that I do not think I would have the courage to put the question to him.

In one of the popular recordings made about twenty years ago by "The Weavers," the group sang its song but then fell completely silent, until the leader said: "We will now sing the same song again—this time, louder." This is essentially what Dr. Dawkins has done, in the two books he has published since *The Selfish Gene*.

The later of these books is *The Blind Watchmaker*,[23] which is pitched at about the same semi-popular level as *The Selfish Gene*, and has enjoyed an almost equal success. The earlier one, *The Extended Phenotype*,[24] on the other hand, is a book which probably only a professional biologist could follow in all its details. Still, lay readers can certainly understand enough of it to see that its substance is essentially the same as that of the two more popular books.

To do Dawkins justice, the same song is sung *softer* and better, in one part of *The Extended Phenotype*. This is the general treatment of genetic determinism in Chapter 2, which is distinctly better than the treatment of it implicit in *The Selfish Gene*. Someone had obviously convinced Dawkins, between 1976 and 1982, that causation does not, after all, come in two grades: genetic or "industrial strength" causation, and an inferior everyday non-genetic grade. Dr. Dawkins may reasonably be thought to have learnt this truth at a disproportionate cost to the public, but it is undoubtedly a step in the right direction: that is, away from genetic puppetry theory.

The overall tendency of these two later books, however, is exactly the reverse: they are actually *more* puppetry theoretical than the first one was. We read in *The Extended Phenotype* that "the fundamental truth [is] that an organism is *a tool of* DNA,"[25] and in *The Blind Watchmaker*, that "living organisms *exist for the benefit of* DNA."[26] Such statements abound even more in the later books than they did in the first one. In addition, they are not counterbalanced here, as they were in *The Selfish Gene*, by cheerfully inconsistent statements like the one I quoted earlier: that we have "the power to *defy* the selfish genes of our birth." Far from there being any "with one bound Jack was free" stuff in the later books, genetic puppetry theory, especially in *The Extended Phenotype*, is universal, unrelieved, and carried to the farthest lengths imaginable. It really is, then, "the same song again, this time louder," in these two later books. But alas, the song still makes no more sense than it did at first.

We and all the other organisms "exist for the benefit of DNA," forsooth! It is *impossible* to benefit an H_2O molecule, or an NaCl molecule: that is, a water molecule or a salt molecule. Try it yourself if you don't believe me. Launch a Help a Water Molecule Week and see how you get on. You may well raise some money, but how could you possibly put it to work? Water molecules simply cannot be helped. And no more can DNA molecules—that is, genes—be benefited.

In particular, a molecule of DNA, or of water, or of anything, is *not* benefited by a replica of it brought into existence by this molecule itself, or by something else, or by nothing. However it comes about, the situation is essentially this: there is at one time a certain molecule of M, and at a later time there is M and its replica. Now, what benefit or advantage is there in this change, to anything whatever? M does not benefit by its replica coming into existence: filial piety does not exist among genes. The replica does not benefit by coming into existence. To paraphrase Kant, existence is not a benefit; or, if it is, it is a benefit which can be conferred only on the non-existent. There are no other possible candidates. Hence there is nothing which benefits by this change, or is better off at the later time than it was at the earlier.

It is true, of course, that if M is a gene, and brings the replica into existence (and survives this process), then there is a larger number of this kind of gene in existence at the later time than there was at the earlier. But *this* proposition implies nothing whatever about benefit. Indeed, it is not even a truth of biology; it is only the trivial truth of arithmetic, that two is a larger number than one. It is equally true that if M is a water molecule, and remains in existence while its replica is synthesized in some laboratory, then there is a larger number of *that* kind of molecule in existence at the later time than there was at the earlier. But it would be evidently nonsensical, in this case, to speak on anything having *benefited* by the change. And it is no less evidently nonsensical in the case where M is a gene instead of a water molecule, and produces the replica itself.

No, Virginia, you and I are not being manipulated by our selfish genes for their own benefit. There are certain people who are subject to incorrigible delusions of being manipulated, and there are also such things as confidence men. But that is all there is to it: there are no "confidence genes." That class of work calls for both intelligence and purpose, and genes have neither. *They* cannot trick people out of their money by issuing false balance sheets, by writing fraudulent books, or by anything of that kind.

I may be quite wrong, but in reading Dr. Dawkins I have often formed the impression that (in Wittgenstein's phrase) a certain *picture holds him captive*. A picture, namely, of an exceptionally vain author, or parent, or photographer, who delights in surrounding himself with his own writings, or children, or self-portraits. But genes (it can hardly be necessary to say) can no more be vain than they can be selfish. They cannot delight in the number of replicas that they make of themselves. They are not even intelligent enough, after all, to know when they *have* made a replica of themselves.

Essay 8
"He Ain't Heavy, He's my Brother," or Altruism and Shared Genes

. . . we expect to find *that no one is prepared to sacrifice his life for any single person, but that* everyone will sacrifice it *for more than two brothers [or offspring], or four half brothers, or eight first-cousins.*
—W. D. Hamilton, in an article of 1964

I

" ALL COUNTRY PEOPLE hate each other." This is the shocking statement with which William Hazlitt begins his essay on country people.[1] There is some exaggeration in it, obviously. But that, alas, is not what is shocking about it. Rather, what makes it shocking is how little exaggeration there is in it, and how much truth. Now country people are also (as everyone knows) more closely related to one another than city people. And yet sociobiologists believe that the more closely related people are, the more altruistic they are towards one another! This seems an extreme instance of putting difficulties in one's own way.

But then sociobiologists believe, quite generally, that how altruistic any organism is, towards another of the same species, depends on the proportion of its genes which the first shares with the second. This is part of their theory of "inclusive fitness," and is their very favorite idea. You meet with instances of this idea at every turn nowadays, wherever the influence of sociobiology extends: which is to say, in some pretty surprising places. An overseas friend of mine, who is a philosopher and an extremely good one, said in a letter last year, "It is no wonder I love my children: they share half my genes." This was straight out of the sociobiologists' manual. Dawkins says, for example: "It is easy to show that close relatives—kin—have a greater than average chance of sharing genes. It has long been clear that this must be why altruism by parents towards their young is so common. What R. A. Fisher, J. B. S. Haldane and especially W. D. Hamilton realized, was that the same applies to other close relations—brothers and sisters, nephews and nieces, close cousins."[2] The general principle, in Hamilton's own words, was this. "The social behavior of a species evolves in such a way that in each distinct behavior-evoking situation the individual will seem to value his neighbors' fitness against his own according to the coefficients of relationship appropriate to that situation."[3] (That is, according as the "neighbor" is an offspring, a sibling, a cousin, or whatever.)

PARENTS DO, OF COURSE, share half their genes with each offspring (as even we laymen now know); grandparents share one-quarter with each grandchild; siblings share (on the average) half of their genes with each other; cousins one-eighth; and so on. All of these are indisputable facts. What, by contrast, is not at all indisputable is the theory that the degree of altruism which exists between a person and his or her relatives corresponds to, and depends upon, the proportion of his or her genes which this person shares with those relatives. That is the distinctively sociobiological idea. You can find it stated, or implied as part of the theory of inclusive fitness, in many places. For example, R. D. Alexander's *Darwinism and Human Affairs* (especially Chapter 10); Dawkins's *The Selfish Gene* (especially Chapter 6); M. Ruse's *Taking Darwin Seriously*; R. Trivers's *Social Evolution*, (especially Chapters 6–8);[4] and so on.

I do not doubt that there is *some* connection between the degree of our relatedness to other people, and the degree of our altruism towards them. But then, there is some connection between, for example, Newton's Laws of motion and the present state of the solar system; in fact a great deal of connection. Yet it would be manifestly silly to try to explain the present state of the solar system just by those laws. You would obviously need a great many other propositions as well. And likewise, you would obviously need a great many other facts, beside the degree of relatedness between two people, in order to explain the degree of altruism, if any, which exists between them.

And then, as we all know, there is some connection between far too many pairs of things for the mere existence of "some connection" to be at all interesting. There is some connection between being fond of pastry and being of Cornish descent, between keeping a pet and being an alcoholism risk, and so on, forever. So, while it is not saying absolutely nothing, it is saying extremely little to say that there is some connection between the proportion of genes we share with our kin, and the degree of our altruism towards them.

Parental altruism is, of course, the strongest as well as the most universal form of kin altruism. As a general rule, that is. It is certainly not so very strong and universal as to prevent, for example, many women being more devoted to their horses, dogs, or cats, with whom they cannot possibly share any interesting proportion of their genes, than they are to their children, at least once the period of their infancy is over.

And then, on the other side, sociobiologists would have us believe that there is no altruism, or none to speak of, outside kin altruism; but the falsity of this is very obvious. In fact altruism, in its strongest and most life-consuming forms, is not directed towards relatives at all. Think of people like Mother Teresa, Florence Nightingale, Father Damien, and Albert Schweitzer. It is of course fashionable, and it is eminently sociobiological, to be impertinent about such people, and I do not doubt that vanity and self-deception could be shown to have found ample outlets in their lives. So they do in every human life: that should go without saying. But it would require more impertinence than I at least can muster to believe that those four people were not more altruistic than the average parent. And if your biology makes a "problem" out of the very existence of such people—as sociobiology certainly does—then what that shows is just that there is something wrong with *your biology*: not that there is something wrong with those people.

II

Altruism ought to be non-existent, or short-lived whenever it does occur, if the Darwinian theory of evolution is true. By the very meaning of the word, altruism is an attribute which disposes its possessor to put the interests of others before its own. Disposes it, for example, to defend conspecifics in danger, when it could have simply saved its own skin, disposes it to eat less, or less well, or later, if this helps others to eat more or better or earlier; disposes it to mate later or less often, if this helps others to mate sooner or more often; and so on. But any such behavior by an organism clearly tends to lessen its own chances of surviving and reproducing, and altruism is therefore an attribute which is injurious to its possessor in the struggle for life. And in that struggle, Darwin says, "we may feel sure that any variation in the least degree injurious would be rigidly destroyed."[5]

But in fact, obviously, altruism is not "rigidly destroyed." On the contrary, it is common in the animal world, at least in its parental form. This is, in essence, the famous "problem of altruism" which has always beset Darwinism.

This problem is evidently a self-inflicted injury, and as such deserves no sympathy. It is just like the even more famous "problem of evil" which has always beset Christian theism. If you don't believe the theory that God exists, and is perfectly good and omnipotent, where is the problem in the fact that evil exists? There is none. If you don't believe the theory that conspecifics are always struggling for life with one another, where is the problem in the fact that altruism survives? There is none.

Organisms would not struggle for life with one another, of course, if they were indifferent about their own survival and reproduction, or if they positively inclined to the Buddhist side of the question, and actually preferred death, and leaving no or few descendants, to the opposite things. But that is not, to put it mildly, the way organisms in general are. Suicide, voluntary sexual abstinence, and contraception are exceptional even in our species, but they occur nowhere else at all. The general rule is that organisms act in a way which tends to increase or maintain their own chances of surviving and reproducing, never in a way which tends to decrease them.

That, at any rate, is what the Darwinism of the nineteenth century said, and what Darwinism continued to say up to the mid-1960s: that organisms behave in general in a way which maximizes their individual fitness. And this is why the element of unselfishness or self-subordination, which is manifest in parental altruism, presented the theory with the serious and widespread problem that it did.

A less starkly individualistic version of Darwinism—the theory of inclusive fitness—was put forward by W. D. Hamilton in 1964,[6] though J. B. S. Haldane and R. A. Fisher, decades earlier, had several times stated the germ of the theory.[7] Its general idea is as follows. An organism acts in such a way as to maximize, not its individual fitness or chances of surviving and reproducing, but its inclusive fitness: that is, the fitnesses of a group of conspecifics which includes, first, the organism itself, then those with which the organism shares the highest proportion of its genes, then those with whom it shares the next highest proportion of its genes, and so on.

The inclusive fitness of a given organism is thus the aggregate or sum of a number of "individual fitnesses." The fitness of the organism itself will always be the largest single component of this sum. If we give to this component the numerical value one, then the component contributed by one offspring of that organism will be exactly one half: the component contributed by one sibling will be about one half; the component contributed by one grandchild will be one-quarter; and so on. But in fact, of course, an organism will hardly ever have only one offspring, or only one sibling; and the fitnesses of two offspring together contribute as much to an organism's inclusive fitness as the fitness of the organism itself does. Three offspring, or three siblings, together contribute a component of one-and-a-half units: half as much again as the organism itself. Two offspring, each of which itself had two offspring, would altogether contribute twice as much to inclusive fitness as the organism itself. Etc.

This theory, as will be obvious, still accords a certain unique position to selfishness in the life of each organism. But it will be equally obvious that, unlike pre-Hamiltonian Darwinism, it also leaves open the possibility of altruism, or at any rate of altruism towards close relatives. Indeed, it does much more than that. For the theory positively predicts that kin altruism will not only exist, but be common, and strong.

Suppose that an animal, at a given moment, must do one or other of two things. One is to let three of its offspring be killed by a predator, while it saves its own life. The other is to lose its own life in saving the lives of all three offspring. If the organism does the former, the result will be a net loss to its inclusive fitness of one-and-a-half units. If it does the latter, the result will be a loss in inclusive fitness of only one unit. The theory says that organisms act in such a way as to maximize their inclusive fitness, hence that they will prefer the smaller of two alternative net losses. So it leads us to expect that this organism will in fact do the altruistic thing, rather than the selfish one.

Between an organism and one other to which the first is related, the theory predicts that the degree of altruism will depend just on the degree of relatedness: that is, on the proportion of its genes which the first organism shares with the second. But of course the *general* theory of inclusive fitness is addressed to the far more usual and important case: the one-*many* case. There, the number of the "many," and not merely their degree of relatedness to the "one," comes into the calculation of an organism's inclusive fitness.

WHAT I HAVE just given, I need hardly say, is only the barest possible outline of the theory of inclusive fitness. As Hamilton first formulated it, the theory was both formidably mathematical and hedged with many biological qualifications. No doubt, between 1964 and now, the mathematics of it have become even less accessible to mere laymen, and the biological qualifications have become even more numerous. But the theory, almost from the moment it was first published, began to revolutionize evolutionary biology, at least at its most general level; and even a layman can sufficiently see why.

The reason was that the theory suggested an explanation of various facts which had previously been anomalies for Darwinism, because they involve attributes which are injurious to the organisms that possess them. One such fact was senescence: why do hereditary infirmities accumulate in age and culminate in death? Another was small clutch size in birds: why do swifts, for example, who could easily lay more eggs, lay only one a year? A third was the combination of bright colors with distastefulness in many species preyed on by birds: a tasted caterpillar is a dead caterpillar after all, while bright colors *attract* predators.[8] The general idea of the explanation in each case was the one which Fisher and Haldane had suggested thirty years before Hamilton's article. Namely that an attribute, which is injurious to one organism that possesses it, could survive in a population by assisting the survival and reproduction of many close relatives of that organism: that is, of individuals which are more likely than others to share the gene which is the basis of that attribute. What Hamilton's theory did was to give that general idea a detailed and quantitatively definite form.

The most striking success of the theory, however, was in explaining another attribute injurious to its possessors: altruism, or at any rate, one form of kin altruism. As is well known, many of the social hymenoptera (ants, bees, wasps, etc.) possess a class or "caste" of sterile workers. The members of such castes are females who do not themselves reproduce but pass their lives in assisting the reproduction of their mother the queen, by looking after their younger siblings. How anomalous such lives are, from the point of view of individualistic Darwinism, will be obvious. For these workers seem to be engaged in maximizing not their own chances of survival and reproduction, but those of another individual, the queen. This anomaly was certainly obvious to Darwin: he tells us that the existence of these sterile castes had seemed to him, at one stage, a fatal objection to his theory.[9]

From the point of view of *inclusive* fitness, however, the existence of sterile workers is much more intelligible. For in these species (owing to an unusual feature of the male reproductive cells), sisters share with each other three-quarters of their genes, instead of one-half as in all other sexually reproducing species; while mothers share with their daughters only the regulation one-half of their genes. A daughter of the queen is therefore more closely related to any one of her sisters than she would be to any daughter that she herself might have. Her inclusive fitness, consequently, can be enhanced *more* by her caring for a sister than it would be by her caring for a daughter of her own.

This is easily the most arresting of the explanatory successes which the inclusive fitness theory has so far enjoyed. But there are by now many other contexts in which the theory has been found to cast at least some explanatory light. Well, there *ought* to be: it has certainly been tried often enough. Inclusive fitness, or some modification of it, has in fact become "all the rage" in evolutionary biology during the last thirty years. A representative recent textbook, such as Trivers's *Social Evolution* (1985), is entirely dominated by it.

Inclusive fitness theory, though thoroughly in the spirit of Darwin, is unquestionably an addition to the older Darwinism. Darwin, like nearly everyone else at the time, had been completely ignorant of Mendel's discoveries when he died in 1882. But even after Mendel's work was rediscovered in 1900, Darwinism remained, for several decades, neglectful of what might be called the "horizontal" line in Mendelian inheritance, as distinct from the "perpendicular" one: that is, the line which links an organism, not to its parents and its offspring, but to its siblings, cousins, and so on. Once, however, Mendel's discoveries had been thoroughly absorbed by the minds of Darwinians—as they were by the generation of Fisher and Haldane—it was only a matter of time before someone said what Hamilton did say in 1964. Namely that, after all, "there is nothing special about the parent-offspring relationship except its close degree and a certain fundamental asymmetry. The full-sibling relationship is just as close."[10]

A natural enough thought, in retrospect. Yet it was sufficient to draw attention to a dimension of inheritance and evolution which had previously been neglected; and sufficient, once developed into a positive theory, to give Darwinism some addition to its explanatory power.

At least, that is what nearly all evolutionary biologists nowadays think. I have written so far in this section as though I think so too; but that was merely in order to avoid mixing up exposition of the inclusive fitness theory with any expression of disagreement with it. In fact I do not believe that that theory *does* explain, or even help to explain, anything. For to say that a theory explains or helps to explain something implies that it is true, or is at least a close approximation to the truth. Whereas I do not believe that the theory of inclusive fitness is true, or anywhere near the truth. My reasons are given in the next section.

But the name "the theory of inclusive fitness" is a peculiarly inexpressive one. Another name for it that is often used, "the theory of kin selection," is positively misleading. For these reasons, I will in what follows sometimes refer instead to the theory as "the shared genes theory of kin altruism." This is cumbrous, but it will at least remind us of what the theory says.

III

NO ONE BELIEVES that Sydney's noon temperature each day, or on a given day, depends on its latitude. The reason is obvious: that Sydney's latitude is constant, whereas its noon temperature varies greatly from day to day. In the same way, no one ought to believe that parental altruism in our species, or in any given sexually reproducing species, depends on each parent's sharing half of his or her genes with each offspring. The reason is obvious. Namely, that this characteristic—each parent sharing half of its genes with each offspring—is common to virtually all sexually reproducing species whatever, whereas parental altruism varies in these species as widely as it *can* vary. Namely, from zero, in all plants and many animals, through countless intermediate degrees, up to its highest degree in the case of man.

This objection to the shared genes theory of kin altruism is so extremely obvious that, when it first occurred to me, I felt sure it must rest on some misunderstanding on my part. So when my friend said in his letter that he loved his children because they share half his genes, I wrote back that at that rate pines and cod would love their offspring as much as we love ours; which they do not. I was hoping and expecting to learn, from his response, what my misunderstanding of the theory was. But I did not. Since then I have put the same point to several other friends who are favorably inclined toward sociobiology, but still without any enlightening result. Yet Dawkins says (as we have seen) that "it has long been clear" that the proportion of genes which parents share with their offspring "must be *why*" parental altruism is "so common"; and what sociobiologist disagrees with that? Until, therefore, someone will condescend to make me better informed, I must continue to think it is a good objection to the theory, that pines and cod and in fact most sexually reproducing species, although they share the same proportion of their genes with offspring as we do, do not come anywhere near us as far as parental altruism is concerned.

ACCORDING TO the shared genes theory of kin altruism, the helpfulness of human brothers and sisters towards one another is due to their having (on the average) half of their genes in common; just as the helpfulness of human parents towards their children is supposed to be due to their having exactly half their genes in common. If this is true, then what vast quantities of altruism must exist, between generations or between siblings, in all those species which reproduce either parthenogenically or by fission! For the members of these species share *all* their genes with their offspring or with their siblings.

If there is one thing which dignifies common human life, and goes some way to relieve its overall charmlessness, it is parental love. Yet human parents share with their offspring only half their genes. Imagine then, if you can, the perfect altruism which a parthenogenic offspring must receive from *its* parent! This probably explains why that chap Jesus had such a vast idea of his own importance. And every time a bacterium divides into two genetically identical "daughter" bacteria, what complete and selfless altruism must unite those two sisters! The contrast with our own meager efforts in the way of sibling altruism is too glaring for any of us to contemplate without pain and mortification. Sociobiologists deserve our thanks, then, for reviving the pre-Darwinian tradition of Mrs. Gatty, by drawing edifying *Parables from Nature* (1855). Yes, even from the meanest of our fellow creatures.

I can see only one difficulty with this uplifting prospect that sociobiology opens up. This is that, like every other uplifting prospect, it is inconsistent with the Darwinian theory of evolution. In species which reproduce parthenogenically (like many dandelions) or reproduce by fission (like bacteria), what will be left of the famous Darwinian "struggle for life?" Bacteria and dandelions certainly fulfill the *Malthusian* part of the conditions required for that struggle. If there are any organisms which you can safely rely on to multiply with maximum speed up to the number that there is food to support, bacteria and dandelions are among those organisms. But once they have got to the Malthusian limit, what then? Two bacteria of the same parentage have 100 percent of their genes in common, and therefore must, according to sociobiology, exercise 100 percent altruism towards each other. So how are *they* going to be able to compete with one another for "the means of subsistence?" Why, they would never even be able to decide which one of them was to go through a doorway first. Yet if no struggle for life, then no natural selection; and if no natural selection, then no evolution. That is what the Darwinian theory used to say, anyway; still does say, come to that. So how is the theory of inclusive fitness to be reconciled with Darwinism?

On second thoughts that is *not* the only objection which asexual reproducers, such as bacteria and dandelions, present to the inclusive fitness theory. As well as that theory's inconsistency with Darwinism, there is another objection: its inconsistency with the facts. For between sister bacteria, and between parthenogenically reproducing dandelions and their offspring, there is *no* kin altruism. Not just much less of it than the theory leads one to expect: there is none at all. Two sister bacteria, despite their genetic identity, will slug it out with each other for the means of subsistence, just like any other pair of good Darwinian girls.

A fact so awkward for the inclusive fitness theory, and at the same time so obvious, was bound to demand attention from the first. Accordingly, Professor W. D. Hamilton, in the famous article in which he first put forward the inclusive fitness theory, tried to apply a patch to this obvious puncture. But I cannot report that he succeeded. He first says that the extent of asexual reproduction may have been greatly exaggerated: which is understandable enough, in both senses of "understandable." After that point, however, I at any rate am unable to understand what the patch is which he proposed to apply; and still less able, therefore, to tell whether it is a good one.[11]

PART OF WHAT the inclusive fitness theory says is that people love their children because they have half their genes in common with each child. But children are not the only things which people have half their genes in common with. Each woman shares half her genes with each egg she produces, whether fertilized or not, and a man shares half his genes with each sperm he produces. The theory of inclusive fitness would therefore seem to predict that every woman loves each of her eggs as much as her children, and that each man loves every one of his sperm like a father his son. This certainly does not sound like anyone I know: not even remotely like. But I cannot deny that my experience of life is small. Does it sound like anyone you know, or have ever heard of?

And if altruism is proportioned to shared genes, what about the converse case? Your wife shares only half her genes with each egg, and you share only half your genes with each of your sperm. But an egg of your wife, and a sperm of yours, has *all* its genes in common with the adult organism which produces it. So if what causes altruism is shared genes, our eggs and sperm must be putting in a 100 percent altruistic effort towards us. Yet as far as I know, nothing of this kind has ever been observed. Parents nowadays often complain, it is true, that they have produced a number of affectionate little layabouts who can scarcely be prevailed on to leave home. But I have never heard of this complaint being levelled against either eggs or sperm, and it would seem to be a complaint entirely without foundation in either of those cases. Indeed, in the case of sperm, it would surely be the reverse of the truth, since they are, if anything, in culpable haste to leave home.

Then, if human brothers have the amount of mutual helpfulness that they generally do, because they share about half of their genes, two sperm of any one man ought to exhibit the same degree of mutual altruism. For they too share about half of their genes with each other. Can you believe this? I can't. I cannot believe even that two sperm of one man have either liberty or equality, but fraternity seems to me entirely out of the question. Even from the point of view of sociobiology itself, the idea is absurd. A sperm is just a packet of paternal genes, after all, and according to sociobiologists, every gene is ruthlessly selfish. So how come, if every gene in every packet is selfish, that *the packets* are so all-fired fraternal? In any case, if what some other biologists report is true, the relations among the sperm of any one man are in general rather the reverse of fraternal.

IF THE ALTRUISM of parents towards their offspring is due to their sharing half of their genes with each offspring, then filial altruism ought to be as common and strong as parental altruism. For if your offspring has half of your genes, then it is also true that you have half of your offspring's genes. Yet in our own species, as everyone knows, parental altruism vastly exceeds filial, both in commonness and in strength. And in the great majority of sexually reproducing species, if there is any parental altruism at all, filial altruism is, by contrast, even more conspicuous by its absence or comparative weakness than it is in man.

In his basic article Hamilton referred, as we have seen, to the fact that the parent-offspring relationship is itself asymmetrical. (See the text to Note 10 above.) But he nowhere refers in that long article to the asymmetry between the *altruism* of parents towards offspring, and that of offspring towards parents. And in fact, the general principle which he enunciated (see the text to Note 3 above) actually requires filial altruism to be equal to parental. For it says that degree of altruism varies according to degree of relatedness; and the degree of relatedness of child to parent is the same as that of parent to child.

I am completely unable to explain Hamilton's silence about the universal asymmetry, where his theory required symmetry, between filial and parental altruism in sexually reproducing species. This was (one would have thought) a puncture in his theory even more conspicuous than the one about the lack of kin altruism among asexual reproducers: which puncture he had *attempted*, at least, to patch. But whatever the reason for it may have been, the fact is that he did not attempt any kind of repair with respect to the symmetry of parental and filial altruism.

A patch was not long in forthcoming, however. The theory which we met with in the preceding Essay, that genes are the only causal agents intrinsic to organic affairs, and are selfish, supplied the needed repair. That theory was a natural outgrowth of the theory of inclusive fitness, as sociobiologists have always acknowledged, and as will be obvious, in retrospect, even to a layman if he reads Hamilton's 1964 article.[12] The patch goes as follows. A parent is necessarily older than its offspring, right? An offspring therefore has more of its reproductive career ahead of it than a parent of it has, right? So a selfish gene, always on the lookout to maximize the representation of its copies in the population, will in general prefer to invest in an offspring rather than in its parent, and will dispose an organism which carries it to honor its sons and daughters rather than its father or mother.[13] (Nothing to it really, once you learn to think of genes as investors who are about a million times better at their work than even the cleverest of human investors. See Essay 9 below.)

But even if we regard this as an acceptable patch for the filial altruism puncture, the same problem breaks out again elsewhere, and in an even more hideous form. Namely, in the form mentioned a moment ago: that sperm and eggs ought to exhibit "filial" altruism towards the organisms that make them, and in fact twice as much altruism as they receive from those organisms. As far as I know, no attempt has ever been made to patch *this* puncture. Perhaps it was considered that even the best patch would be bound to draw attention to it.

THE SHARED GENES theory of kin altruism suffers from other punctures which no attempt has ever been made to patch, for the simple reason that they are, in the eyes of the theory's adherents, not punctures at all, but beauty spots. One of these concerns identical twins. Such twins have, of course, all of their genes in common. Their mutual altruism must therefore, according to inclusive fitness theory, be 100 percent.

This *reductio ad absurdum* of the theory is willingly embraced, in fact mistaken for a successful prediction, by all sociobiologists. And not only by them. Professor G. C. Williams, for example, writes: "To provide benefits to a genetically identical individual is to benefit oneself."[14]

So it is crystal clear what any young man A ought to do, in the interests of maximizing his inclusive fitness, if he is lucky enough to have an identical twin B. He should propose that B refrain from having any children of his own, and instead devote all his money, energy, and indeed his life, to maximizing the number and fitness of A's descendants. This proposal is sure to sound an extraordinarily selfish one, to persons unversed in the theory of inclusive fitness. But adepts in that theory are well aware that selfishness is *impossible* between identical twins. And after all, if Professor Williams's statement is true, what else could B possibly do, that would be more beneficial to his own inclusive fitness than this course of action proposed by A? In fact, if B does not fall in with A's proposal, it can only be a case of "biological error,"[15] or to speak plainly, stupidity.

THERE ARE YET other untoward consequences of the theory of inclusive fitness, which have neither been recognized as requiring a patch, nor been mistaken for beauty spots. They are simply never (at least as far as I know) mentioned at all. One of these concerns the advantages which can be gained from incest, both between siblings, and between parents and offspring, if the theory is true.

There is, as everyone knows, all too much conflict and disharmony in human life, much of it directly attributable to the weakness of our altruistic tendencies; or in plain English, to selfishness. A particular *locus* which is notorious for disharmony is where three generations of one family live in close proximity. Now, this is an area where, simply by applying the theory of inclusive fitness, it would be not only possible but easy to bring about a great lessening of disharmony.

The simple application that I have in mind is as follows. You should first marry one of your siblings, and then make sure that you or your spouse are a parent of any children that your children have. This way, you will have three-quarters of your genes in common with each of your children, and nearly nine-tenths in common with each grandchild. The resulting all round increase in kin altruism, or decrease in disharmony, is bound to be very great, if the shared genes theory is true.

I am not, of course, either recommending this policy myself, or suggesting that anyone else either does or should recommend it. I merely point out that it would be an easy and effective way, if the shared genes theory is true, of greatly increasing kin altruism. The policy would, quite obviously, be attended by a certain danger: that of encouraging expression in the family phenotype of harmful genes which would remain recessive under domestic arrangements of a more conventional kind. In any case, reduction of disharmony, while highly desirable, is not the only desirable thing in human life, and is not necessarily a goal to be pursued regardless of cost.

But suppose that the reduction of disharmony *were* the only or the supreme goal: it would still be far from certain that the policy I have described would achieve that end. It *would*, indeed, if the shared genes theory of kin altruism is true. But common sense suggests that it would not. For a pronounced bias *against* incest is discernible almost everywhere in animal and even plant life. To fly in the face of that bias, in the way I have described, would be more likely to produce, in fact, a great *increase* in family disharmony, simply because of the aversion to incest.

But who would ever suspect, from the shared genes theory of kin altruism, that there is, or even that there could be, such a thing as the aversion to incest? There is not one human being who does not prefer to be on the receiving end of kin altruism, rather than of either kin hostility or kin indifference. There is not one of us who does not feel that he could use rather more kin altruism than he actually receives. I have simply pointed out an easy way in which, if the theory of inclusive fitness is true, we could all *get* more.

CUCKOLDED HUSBANDS who are also fathers, and do not know they have been cuckolded, are not noticeably fonder of the children who really are their own, than of those they mistakenly believe to be their own. When a couple adopts a baby, they are not as a rule less fond of it than parents are of their own babies. But rather than run through all the obvious objections of this kind to the inclusive fitness theory, let us consider what would be the effects of a worldwide, simultaneous, and unsuspected "baby switch."

It sometimes happens in maternity hospitals that a baby is given to the wrong mother, and that the mistake goes undetected for some time. Well, let us suppose that on a certain day, every child who is born anywhere in the world is somehow (it does not matter how) given to the wrong mother, and that no one even suspects that this has happened, or ever will suspect it. What will be the effect of this switch, on the altruism of the parents towards "their" babies who are born on this day?

The theory of inclusive fitness is a causal theory about kin altruism: an attempt to say what causes it, or what kin altruism causally depends upon. That is, it is a proposition of the same general kind of the theory (for example) that the tides causally depend on the moon's gravitational attraction, or the theory that polio myelitis depends on such and such a virus. And what the theory says that kin altruism depends upon is the degree of relatedness between the organisms in question. More specifically, the theory says that the altruism of parents towards their offspring depends on their sharing half their genes with them.

On the day of the universal baby switch, therefore, the thing which, according to the theory, parental altruism depends upon is altogether missing. The parents, and the baby that they take home, are simply not related to one another in the way they think they are, and indeed, except *per accidens* are not related to one another at all. Every one of those babies, consequently, is going to feel the effects of a total absence of parental altruism towards it. There is going to be, in fact, a simultaneous worldwide disappearance of parental altruism. Infant mortality is going to undergo an enormous and inexplicable increase, etc., etc.

It is likely enough that there is not one person in the world who will actually believe this consequence of the shared genes theory of kin altruism. But there are certainly very many people who, logically, *ought* to believe it. Namely, everyone who believes that theory, and hence believes that the altruism of parents depends on their sharing half their genes with their children.

In general, causal dependence is not "belief sensitive" (as we might put it). The tides depend on the moon's gravity, and polio myelitis depends on a certain virus, absolutely independently of what anyone may know or believe. The tides would still depend on the moon, even if there were not a single organism in the world that was capable of belief at all. There are, however, certain cases in which causal dependence *is* belief sensitive. A stock example is a bank's going broke. This effect can be brought about by any number of different causes, such as the incompetence, or the dishonesty, of its managers. But it can also be brought about by a sufficient number of people *believing* that the bank will go broke, and acting accordingly.

Now the inclusive fitness theory says that kin altruism depends on shared genes, and the causal dependence which is meant here is quite certainly not of the belief-sensitive kind. It had better not be! For that theory is intended, after all, to explain kin altruism, not just in humans, but in any social animals, however low they may be on the intellectual scale. Very many of our dumb friends (we need to remember), if they can be credited with having beliefs at all, cannot possibly be credited with having beliefs about a subject so naturally difficult as their family tree. In other words, the theory of inclusive fitness does *not* say, for example, "You will love the young conspecifics who are in fact your children, as long as you *believe* them to be your children." It does not say anything like that. It says "You will love your children anyway"; just as the theory of the tides which we all accept says that the moon's attraction will produce tides anyway.

It really is a consequence of the inclusive fitness theory, then, that all the babies born on the day of the universal switch are going to be deprived of parental altruism. According to that theory, our parental altruism is as little subject to the control of our *beliefs* as is (for example) the working of our thyroid gland. Quite the contrary, that theory maintains that our parental altruism, our thyroid secretion, and just about everything else about us, is under *strict genetic* control: and under no other control whatever, except that of accidents thrown up by the environment.

It is well known that sociobiologists have darkened immense areas of paper, and imagined they were providing evidence for the inclusive fitness theory, by publishing statistics which show (for example) that people are more likely to mistreat a child they have adopted, or carried over from a previous marriage, than a child of their own current marriage. This sort of thing is, in fact, a major division of the sociobiological industry. But it would not be easy to conceive a more pointless expenditure of effort.

For one thing, it is completely unnecessary, because we knew it all long ago. Traditions about cruel stepmothers, the misfortunes of foundlings and the like, are universal and ancient, and no one has ever thought that they are altogether without a good deal of foundation in fact. Parents, like all other humans, are exceedingly imperfect, and some of them really are as bad as adolescent children often imagine them to be. Everyone knows that. But everyone also knows, and always has known, that a child who goes further than its parental home will, in all probability, fare even worse.

But the sociobiologists' statistics about the probability of an adopted child being mistreated, etc., are not only unnecessary: they are completely worthless as evidence for the shared genes theory of kin altruism. The reason is that they are subject to an enormously high level of what experimental scientists call "noise." That is, it is quite impossible to determine how far the observed effects are due to the cause which is being "tested for"—in this case, the actual degree of relatedness—and how far they are due to other causes altogether.

The main source of this noise, in the case of adopted children or stepchildren, is of course the fact that such children are *known* to be so, by their adoptive parents or stepparents. Adopting a child, or having a stepchild, is not, after all, something you can do in your sleep, or while in a coma. And then, humans have good memories, in fact the best—by about a billion miles—that there are. So human couples know, and do not forget, that they stand in a different biological relation to an adopted child or a stepchild, from the relation they stand in to any fruit of their own union. (Of course they need not know anything whatever about *genes*.)

And the consequence is that when a couple treats an adopted child or a stepchild worse than they treat their own children, it is utterly beyond the power of any mere *statistical* analysis to reveal how far this difference depends upon biological fact, and how far it depends on "noise" created by what the couple know or believe. How far, that is, it depends on a child's *being* a stepchild or adopted, and how far it depends on its being *known* to be so. Much of what people know or believe, after all, is *not* under genetic control at all: it can easily depend (for example) on which newspaper you happen to pick up. And even the most rabid sociobiologists do not think that which newspaper you pick up is under strict genetic control; although, at the present rate of intellectual progress, in a few years' time some of them *will* think this.

Anyone who possesses a spark of methodological morality (as sociobiologists do not) can easily see what kind of demographic information really would—if only it could be got—be evidence for the shared genes theory of kin altruism. It would be, for example, a couple who have children of their own, adopt two others, one of whom happens to be (though they do not know it) a child of their own whom they had earlier abandoned or "adopted out," and who treat *this* adopted child better than the other, although all other things are equal between the two adoptees. But it can hardly need to be said how rarely such a case must occur, or how unlikely it is, if it did occur, to be accurately reported. Alternatively, unsuspecting cuckolds, who nevertheless discriminate against all and only their suppositious children, *would* be evidence for the shared genes theory. But again, such cases can safely be assumed to be rare, since otherwise the phenomenon would be notorious and even proverbial, like the "cruel stepmothers" of the nursery stories. And the extreme difficulty, or rather virtual impossibility, of obtaining reliable information concerning such a case, will be evident.

IT IS, as the proverb says, a wise child that knows its own father.

But in this respect there is (to echo Hamilton's famous remark) "nothing special" about a child's ignorance of its *paternity*. It is an equally wise child who knows its own mother. It is a still wiser one who knows which people are its siblings, its aunts, its cousins (etc.). Children are *told* these things, of course, and they believe them simply for that reason. But believing things just because you are told them is not knowing them. You recognized your "mother's" face (as is by now well known) only a few days after your birth. But then, you would have done exactly the same, if you had been one of the babies born on the evil day of the universal switch. Which shows how much, in real cognitive terms, your "mother recognition" is worth: namely, nothing at all.

Now, according to the shared genes theory, an organism distributes its degrees of altruism towards surrounding conspecifics according to the degree of relatedness in which it stands to those conspecifics. Except in populations of exceptionally high "viscosity" (as Hamilton says)—that is, where the individuals "get around" extremely little—this discriminating distribution of altruism plainly requires that an organism be able to distinguish both between conspecifics to whom it is related and those to whom it is not, and, among the former, between those to whom it is most closely related, and those to whom it is less so.

As far as the parental recognition of "own young" is concerned, the theory stands up passably well. Or anyway, half of it does. This is, the half which concerns *maternal* recognition of own young. According to the strict letter of the theory, of course, paternal recognition of own young ought to be as common and unerring as maternal recognition, since the genetic contribution of fathers to offspring is equal to that of mothers. But, well, . . . let it pass. In fact, as everyone knows, fathers in the great majority of species are decidedly vaguer than mothers as to which young are their own. And the reasons for this are quite obvious enough, without the lengthy exposition which sociobiologists love to lavish upon them.

The inclusive fitness theory, however, implies far more kin

recognition than just parental recognition of young. At least in its 1964 version, the theory implies that the converse ability, of young to recognize their own parents, is as common and unerring as their parents' ability to recognize them. Even in its present-day version, the theory implies that the ability of siblings to recognize one another is very nearly as common and unerring as the ability of parents to recognize their offspring. It likewise implies that a social animal is able to tell the difference between a sibling and a cousin. For Hamilton's fundamental thought (as we have seen) was that "there is nothing *special* about the parent-offspring relationship . . . the full-sibling relationship is just as close," while the cousinship relation is only a lesser grade of the same thing. That, after all, was the whole idea.

Yet it is perfectly obvious that every one of these implications of the theory is false. The ability of young to tell their own parents from other adults is nothing like as common and unerring as the ability of parents (or anyway mothers) to tell their own young from others. (Since young *care* little who feeds and protects them, as long as someone does, while parents care much whom they feed and protect, it would be surprising if these converse abilities *were* anywhere near equal.) The ability of siblings to recognize one another is not a hundredth part as common as the ability of parents to recognize their young. And the ability to recognize a cousin, as like a sibling only less so, is far rarer still. A cousin, in the vast majority of animals, is just another stranger conspecific.

These facts are all obvious enough, and it would clearly be asking too much of animals in general to expect them to be otherwise than they are. The intellectual capacities of animals are just hopelessly unequal, both in extent of view and in fineness of grain, to the demands which the theory of inclusive fitness makes upon them. It is true that that theory has led to the discovery of a few previously unsuspected cases of kin recognition, especially among some of the social insects. But overall, the fact is that there is among animals only a tiny fraction of the amount of kin recognition that the theory implies.

Indeed, even the ability to recognize a *conspecific* is nowhere near as common and unerring as natural selection, if it is as powerful a force as Darwinians think it is, might reasonably be expected to have made it. Male insects of many species "copulate" with a part of a flower which mimics a female conspecific of the insect. When defending a territory, a male robin red breast (as some famous experiments showed) cannot even tell the different between a trespassing rival male, and a bit of red wool on a wire.

Nor can it be maintained that "in the wild" it is only mistakes of the *opposite* kind that would matter: that is failures to recognize a conspecific *as* a conspecific. That is very far from being the case. The nest parasitism of cuckoos, for example, is the huge success that it is precisely because parents of the host species cannot recognize a non-conspecific's egg *as* the egg of a non-conspecific. Species of ants, which are parasitized by other species, are further examples of the same kind. In fact, parasites and predators are always waiting to take advantage of *either* kind of intellectual defect: inability to recognize a non-conspecific *as* a non-conspecific, or inability to recognize a conspecific as a conspecific.

A male robin red breast then, at least when defending a territory, cannot even tell the difference between a bit of red wool and a trespassing rival, even though a trespassing rival could quite easily be his brother. Yet the theory of inclusive fitness requires us to believe that he can tell the difference between his brother and a cousin, and again, between a cousin and an unrelated conspecific. Well, it is not logically, or even biologically, impossible. It is just incomparably more probable that he cannot.

If he *did* have all this cognitive ability, what use could he possibly make of it? A robin defending his territory is an extremely busy man, and he is not running a charity either. So if this bum turns up at the front boundary and claims to be his long-lost cousin or something, and the robin can tell that he is too, and can even tell that the fellow is in fact his brother, what can he *do* with this knowledge? Even if he is as altruistically disposed to the bum as the shared genes theory says he must be, he *cannot* get him a paid job, or find him a wife who works or is rich. What can he do about the situation at all? Nothing whatever. So all the genealogical knowledge, which the theory of inclusive fitness credits him with, will be perfectly useless, both to himself and to his brother. The only thing he can do is what he will do: namely, tell his brother to clear off, exactly as he would tell a complete stranger.

In genetic terms, of course, the "horizontal" relation of siblings to each other is almost as close as the "perpendicular" one of parents to offspring. The inclusive fitness theory says that degree of kin altruism depends on degree of genetic relatedness. So the theory says that sibling altruism is about as common and strong as parental altruism.

And what can one possibly say, in response to a falsity as breathtakingly obvious as this, except that it *is* false? Parental altruism is so very common, and is commonly so very strong that it really is (as I said in Section II above) a problem for Darwinism. But no Darwinian has ever lost, or should have lost, a minute's sleep over *sibling* altruism. There is simply not enough of it about. When and where, among animals, was sibling altruism ever anything to write home about?

Perhaps among those social hymenoptera, the altruism of whose sterile workers it was the greatest apparent triumph of the shared genes theory to explain? Why no, not even there. Those workers do not, to any significant extent, assist the reproduction of their *sisters*; they could not, because the vast majority of these sisters are never going to reproduce, but will be workers themselves. In other words, the workers achieve little, in the way of increasing the number of copies of their own genes among the individuals with whom they have the closest genetic affinity. But on the other hand, they do greatly assist the reproduction of an individual with whom they have less genetic affinity than they have with their sisters: namely their mother, the queen. The altruism of sterile hymenoptera would appear, therefore, to be a case of the hypertrophy of *filial* altruism, rather than of sibling altruism; even though daughters in these species share only the regulation one-half of their genes with their mothers.

But even if (as is likely enough) the preceding paragraph contains some layman's mistake, where has sibling altruism ever been anything to write home about *outside* certain social hymenoptera? Where else has it ever even *distantly approached* parental altruism, either in commonness or strength? The answer is, of course, nowhere.

A puncture on this scale to the theory of inclusive fitness cried out, naturally, for a patch. A proposal for patching it was put forward by Dawkins in 1976, though I do not know how widely his proposal has been adopted. The substance of it was that it is not the *actual* degree of relatedness which determines an animal's degree of altruism: the animal's *estimate of* its degree of relatedness also has a big, or even a bigger, say in the matter. Incredible as it may seem, this really is what he said: that "relatedness may be less important in the evolution of altruism than the best *estimate* of relatedness that animals can get. This fact is probably a key to understanding why parental care is so much more common and devoted than brother/sister altruism in nature . . . ."[16]

This is plainly a proposal for *increasing* the already intolerable demands which the inclusive fitness theory makes on the intellectual powers of animals. An animal is now to be required, not only to have a good idea of its family tree, but to give a *probability weighting* to each degree of relatedness which it assigns to any animal on that tree. This is more than a little unfair to dogs; to say nothing of many of our still dumber friends, such as rabbits and kangaroos.

But Dawkins's proposal also involves an obvious and enormous *departure from* inclusive fitness theory. For that theory implies that parental altruism, sibling altruism, and all the rest, are under strict *genetic* control; but estimates of relatedness are *opinions about details of one's environment*, and such things are not under genetic control. Humans are far better than any other animals at knowing something about their family tree. Yet what was your estimate, that so-and-so is your sister, ever controlled by? By nothing, of course, except having heard your (estimated) mother say things like, "Stop bossing your sister." And even now you might at any moment be required to revise that old estimate, by overhearing a bit of gossip, by reading a newspaper, or by any one of a million non-genetic accidents. Many an adolescent suburban girl's best estimate of her relatedness is that she is of royal birth, and quite unrelated to everyone around her. Is this estimate, or any different estimate that she ever makes, under genetic control? As for dogs, young birds, horses, etc., their best estimates of their relatedness are so little genetically controlled that it is pathetically easy to half convince them that they are *people*. In short, once you allow a sizable causal role, in determining the degree of altruism, to something which is as biologically accidental and variable as *estimates* of relatedness, you have forfeited all claim to be still giving a genetic explanation of kin altruism.

Unless, then, some patch has been found since 1976 which is a great deal better than the one Dawkins then proposed, the theory of inclusive fitness still has the gaping puncture which it had at first. Namely, that it requires sibling altruism to be about as strong and common as parental, whereas in fact it is nothing of the kind. If so, we might as well admit that although, *genetically*, the sibling relation is "just as close" as the parent-offspring relation, *biologically*, it is nowhere near as close, at any rate as far as altruism is concerned. Of course this antithesis, of "genetically" with "biologically," is bound to scandalize all ears accustomed to the theory of inclusive fitness. But we might all just have to be brave about that.

To A *certain* extent, of course, the temperature of Sydney, or of anywhere else on earth, does depend on its latitude. Everyone knows this, because everyone knows that temperature falls off systematically (however irregularly) with increase in latitude. Two places on the same latitude always *would* have the same temperature at any given moment, if all other things were equal between them, and barring accidents.

But then, other things *cannot* ever all be equal between them. Two places on the same latitude must differ in longitude at least, and the angle of incidence of the sun's rays can therefore never be the same at the two places at once. Even the things which could in principle be equal, between two places on the same latitude, never *are* all equal in fact. The two places always differ in some respect which affects local temperature: elevation, or humidity, or prevailing wind, or surrounding topography, or something. But even if, by some unheard-of fluke, the two places were the same in all such respects, meteorological accidents would nearly always make them differ in temperature at any given time. There would be a windy morning here and a still one there, an electrical storm there but not here, or something of that sort.

The result is, what everyone knows: that knowledge of the latitude of a certain place is just about useless for predicting what its temperatures are like, if you happen not to know this, or for explaining the temperatures it experiences, if you do happen to know what these are. There is undoubtedly a causal dependence of temperature on latitude, but it is extremely attenuated. Very many other causes beside latitude contribute to determining a place's temperature at any particular time. And as well as that, any of these causes, and any combination of them, is often prevented, by mere local accidents, from having the effect on temperature which it otherwise would.

The dependence of temperature on latitude, then, though real, is extremely attenuated. But the dependence of parental altruism on parents sharing half their genes with offspring, if it is real at all, must be much more attenuated still. For there is no latitude at which temperatures vary as widely as temperatures on earth *can* vary; whereas (as I said in the first paragraph of this section) species in which parents share half their genes with offspring vary in parental altruism as widely as species *can* vary in that respect.

So the knowledge that in a certain species parents share half their genes with offspring is even *less* use, for either explaining or predicting the degree of parental altruism in that species, than the knowledge of a certain place's latitude is for explaining or predicting its temperature. The statement "Two places on the same latitude would always have the same temperature, if all other things were equal between them, and if no accident made their temperatures differ" is one which says extremely little. The reason is that the two provisos it contains are so stringent that neither of them is ever in fact satisfied. Still, the statement is one which, little as it says, we have very good reason to think true.

Compare it with the statement "Two species, in both of which parents share half their genes with offspring, would always exhibit the same degree of parental altruism, if all other things were equal between them, and if no accident made them differ in altruism." This statement, too, says extremely little, and for the same reason as before: the provisos it contains are too stringent ever to be satisfied. (When are all other things going to be equal between two species in which parents share half their genes with offspring?) But, little as the statement says, what reason have we to think even that little true? We certainly do *not* have in this case, corresponding to what we do have in the case of temperature and latitude, a systematic (even if irregular) decline in degree of parental altruism, with a declining proportion of genes shared by parents with offspring.

Not only do we not have that: we *do* have reasons, several of them, to think that this statement about shared genes and parental altruism is false. Namely, the reasons I have given at various points in the present section. The same thing holds for all the other false consequences which we have found the shared genes theory to have: *none* of them can be saved by a proviso about other things being equal.

In any case, the shared genes theory of kin altruism is *not* put forward, by the writers who think it true, subject to weighty provisos about other things being equal, and accidents aside. At least, I have never met with any exposition of the theory in which it was burdened with either of those dispiriting qualifications; and I have read such expositions by Hamilton, Dawkins, Alexander, Ruse, and Trivers, as well as by many less well-known authors. On the contrary, the theory is always, in my experience, put forward without any qualifications at all. It is flatly stated, in effect, that the degree of kin altruism varies as the proportion of genes shared. In this unqualified form, the theory is indeed, as we have seen, extremely easy to disprove. But that is not because it is a straw man misrepresentation of the real theory. It *is* the real theory.

BUT I HAVE NOT yet mentioned the most obvious of all the punctures that the theory of inclusive fitness suffers from. This is, its prediction that an animal (as long as it is social, and the genetics of its reproduction are of the almost universal kind) will always sacrifice its life to save the lives of three or more conspecifics with each of whom it shares half its genes (such as its offspring or siblings).

There is no question that this prediction *is* a consequence of the inclusive fitness theory: the fact is admitted, or rather, complacently affirmed.[17] For this very conspicuous puncture, Dawkins proposed the same patch as he did for the one about the inferiority of sibling altruism to parental. Namely that if, in defiance of the theory, animals sometimes "value themselves more highly even than several brothers,"[18] the reason must be the difference between their real and their estimated relatedness: that is, these animals just don't realize that those other three are their siblings or offspring. But as we have seen, Dawkins's proposal, as well as being objectionable in itself, amounts to abandoning the theory of inclusive fitness.

Hamilton wrote that "we expect to find that no one is prepared to sacrifice his life for any single person but that *everyone will sacrifice it for more than two brothers [or offspring], or four half-brothers, or eight first-cousins.*"[19] Was an expectation more obviously false than this one ever held (let alone published) by any human being? I do not see what anyone could possibly say in response to it, except that if Hamilton or anyone else really does *expect everyone* to sacrifice themselves for three brothers or three offspring, then his "expecter" is due for a valve job.

It is true I have omitted a qualification which Hamilton prefixed to the words just quoted: namely, ". . . in the world of our model organisms, whose behavior is determined strictly by genotype. . . ." But Professor Hamilton could hardly object to this omission. For his disciples such as Dawkins constantly do the same thing: that is, read off the results of Hamilton's "model" as being true descriptions of biological reality. No doubt the reason is that they believe that the proviso—behavior being determined strictly by genotype—is satisfied everywhere *in fact*. And then, Hamilton has no quarrel with sociobiology.[20]

Even if we confine our attention just to (say) birds, it is manifestly impossible to *generalize* about the readiness of animals to sacrifice themselves for the sake of conspecifics with whom they share half their genes. In some species of birds, parents are daring defenders of their nestlings, in others they are not quite so brave, and in others again they undertake hardly any risks at all for that end. Even within the same species, individual birds differ significantly in their readiness to sacrifice themselves for young. These things being so, what conceivable excuse can there be for saying flatly that a parent will sacrifice its life for three offspring, in all *social species* where parents share half their genes with their young? Yet the theory of inclusive fitness does say exactly that.

IN THE CASES of adoption, and of unsuspected baby switching or cuckoldry, my objection to the inclusive fitness theory was that it predicts too little parental (or rather "parental") altruism. But almost every other objection I have made to the theory was to the effect that it predicts *too much* kin altruism, and in fact *far* more than actually exists.

Parental altruism equal to the human in all species in which parents share half their genes with offspring; parental altruism twice the human in parthenogenic reproducers; sibling altruism twice the human between sister bacteria; parental altruism towards sperm and eggs, twice as much filial altruism *from* them, and sibling altruism among sperm; filial altruism equal to parental; 100 percent altruism between identical twins; greatly increased kin altruism in incestuous families; sibling altruism equal to parental; and parental altruism so strong that a parent would always sacrifice its life to save three offspring.

In fact, if the theory of inclusive fitness were true, the world would be awash with kin altruism, and we would all, from the bacteria up, be swimming (or drowning) in an ocean of love. It would not be "the problem of altruism" which gnaws endlessly at the vitals of Darwinism. It would be "the problem of selfishness"; and the problem would be where to find any of it.

Surely this is a very extraordinary consequence? Such an ocean of love is the very *last* thing that any Darwinian ever meant to imply, and least of all any sociobiologist meant to imply, when he embraced the inclusive fitness theory. Yet it must be presumed that they are at least as competent as other people to see the logical consequences of their theory. Something must have gone badly wrong here, if not with inclusive fitness theory itself, then with my interpretation of it.

It has. In the next section I will explain what it is that has gone wrong.

IV

The philosopher Mary Midgley was not guilty of presumption when she examined, lethally though humanely, what passes for thought about ethics in the mind of E. O. Wilson;[21] nor yet when she took Dawkins to task over "selfish genes" (in an article referred to in the preceding Essay). She may of course have been mistaken in some or all of her criticisms of these sociobiologists; but that is an entirely different thing from it being presumptuous of her, as a non-biologist, to criticize them at all.

I take it that I was similarly not guilty of presumption, in being critical (in Essay 6) of the sociobiologists' selfish theory of human and animal nature, or again in being critical (in Essay 7) of their demonological genetics. The sociobiologists whom I criticize are all professional biologists, whereas I am the merest layman in biology. Still, there cannot be any serious suspicion here of *lèse-majesté*.

It is rather a different matter, though, for a mere layman to venture to criticize the theory of inclusive fitness. It is certainly a good reason for being suspicious of that theory that it gave rise to the ideas about genes which were the subject of the preceding Essay. But on the other hand, the inclusive fitness theory had Fisher and Haldane as its grandfathers, W. D. Hamilton as its father, and (for example) G. C. Williams as one of its first and most enthusiastic adopters.[22] No one who has read these writers will doubt that they are entitled to rank as considerable biological authorities. What importance, then, can possibly attach to objections to the inclusive fitness theory which, like those put forward in the preceding section, are those of a mere layman?

It is logically possible, of course, that every one of my objections is worthless: either mistaken as to fact, or else based on a misunderstanding of the theory that it is intended as an objection to. But as against that possibility, the inclusive fitness theory, in its essentials (as distinct from its mathematical details), is not a hard one even for laymen to understand. Besides, there is *no* real as distinct from logical possibility that I was mistaken in saying (for example) that sibling altruism is not as common or strong as parental altruism. But what mainly reassures me that not all my objections can be wide of the mark is the following fact. That at various places where the theory seemed to me, from the time I first vaguely heard of it, to have an obvious puncture—asexual reproducers, sibling altruism, and parental self-sacrifice, for example—a patch *has in fact been proposed*, either by Hamilton himself or by one of his disciples.

I therefore find myself in an impasse. On the one hand, the theory of inclusive fitness has many consequences, the falsity of which is so obvious that even a layman (not to mention an intelligent child of nine) can see it. The falsity of these consequences is therefore, presumably, more obvious still to professional biologists. But on the other hand, the inclusive fitness theory is universally accepted by evolutionary biologists, and indeed is generally regarded by them as the greatest addition to their explanatory power that has been made since the 1930s and '40s, when Darwinism merged with Mendelian genetics. In fact it is understating the case to say that the theory is universally accepted. Inclusive fitness is by now a perfect article of faith with virtually all evolutionary biologists, and with all of the lay readers who take their beliefs about evolution from them. An objection made to the theory is never considered to prove anything except the incompetence of the objector. I did not know this when I first began to express criticisms of the theory. But since that time I have learnt it by experience: as Laplace said in another connection, "*par expériences nombreuses et funestes.*"

Scientists sometimes (as is well known) continue to work with a theory which they themselves know is false. Laymen, when they hear of such a case, are apt to be audibly critical of the scientists' conduct; but of course they have no *better* theory to suggest, and the only result is that the scientists grow angry and impatient with their lay critics. But these features of scientists' behavior are not ones which deserve esteem, and still less, imitation. They are *departures from* rational behavior, not forms of it. They arise only because professional scientists, without the guidance of *some* theory however unsatisfactory, do not know what to do with themselves. But laymen have other occupations, and the indignation they feel, when scientists stick like limpets to a theory they know is false, is not only natural but rational. A rational interest in science, as distinct from a professional one, is an interest in what is true, or probably true, or probably close to the truth: in that, and in nothing else. If a scientific theory is *certainly* not even *near* the truth, then, whatever attractions it may have for scientists, it is of no interest to a person who is simply trying to have rational beliefs and no others. That is how things actually stand, of course, with the theory (for example) that the blood is stationary, or that the earth is shaped like a bullet, or that it rotates from east to west. It is also how things actually stand with the theory of inclusive fitness.

When a proposition is obviously false, and is nevertheless widely and fervently believed, it is a reasonable inference that it possesses some powerful attraction for the minds of those who believe it: powerful enough, anyway, to outweigh its obvious falsity. Take, for example, the theory that human beings are immortal. The falsity of this proposition is obvious now, but it always was as obvious as it now is: it is not as though we have lately discovered the first disproofs of this theory—we have not. Yet it was generally believed in Western Europe for most of two thousand years, and (on the whole) was believed most fervently by precisely the people whose intelligence and education best entitled them to rank as intellectual authorities. What the attraction of the theory was in this case is too obvious to need stating.

The theory of inclusive fitness is in an analogous position nowadays, if what I have said about it earlier in this essay is true. That is, it is obviously false, and is nevertheless widely believed, and believed most fervently precisely by the people best entitled to rank as authorities on evolutionary biology. It therefore must possess some powerful attraction for the minds of those who believe it. But what is this attraction?

The answer will easily suggest itself, if we recall certain historical facts. They are all ones which have been stated in Essays 6 or 7 above.

FIRST: ALTRUISM WAS, from the very start, a problem for the Darwinian theory of evolution, if not something worse than a problem. As a result, Darwinians have always been under a certain temptation to "cut the knot," and deny the very existence of altruism. This temptation was always strongest, naturally, for those Darwinians—such as Fisher—who were most convinced of the sufficiency of natural selection to explain everything in evolution. But, of course, for Darwinians to come out and explicitly deny the reality of altruism would have brought Darwinism into open conflict not only with common sense, but with common decency.

For this reason, Darwinians successfully resisted the temptation for a very long time: in fact for more than a hundred years. But after the mid-twentieth century, their views became steadily more and more "selectionist." They were therefore obliged to concede to altruism at least the respectable status of being "a problem" for Darwinism; since it could manifestly not be fobbed off any longer with any designation less than that, or with the old prudent silence. It was even sometimes admitted to be a problem which lies unpleasantly close to the vital parts of Darwinism, and one which at the same time is peculiarly resistant to treatment. But still, no Darwinian yet dared to say, or probably even to think, that altruism is an *illusion*. The uneasy truce on this subject, between Darwinism and common sense, subsisted from 1859 to the second half of the 1960s.

Second: after 1964, this long "gentleman's agreement" was rudely and repeatedly broken, by the Darwinians who soon came to be known as "sociobiologists." No gentlemen Darwinians these, but dragon's teeth, sown by the theory of inclusive fitness, "which sprang up armed men." They adopted, as their principal badge of distinction and the fundamental plank of their platform, the ancient theory of universal selfishness. They proclaimed, as exceptionless biological truths, "dog eat dog," "dirty tricks," "nice guys finish last," the manipulative nature of all communication (etc., etc.), as we saw in Essay 6 above.

Third: sociobiologists are not merely willing, but devoted, "Slaves of the Gene."[23] They believe that an organism—a man, say—is epiphenomenal to his genes: an effect, not a cause. Or at least, they believe that a man is about *as* epiphenomenal to his genes, as his singlet (for example) is to him. Wilson spoke for all sociobiologists, when he said: "An organism is only DNA's way of making more DNA."[24]

Fourth: sociobiologists believe that genes are *selfish*. By which they mean, that everything a gene does is directed to one and only one end: that of maximizing the representation of copies of itself in the next generation of the organisms which carry this gene.

Now that we *have* recalled these four historical facts, however, an interpretation of the inclusive fitness theory, which is very different from the one I have so far taken for granted, will suggest itself to our minds. Or rather, it will force itself irresistibly upon our minds.

I have taken it to be (as I said) a proposition of the same general kind as the theory that the tides causally depend on the moon's attraction, or the theory that polio myelitis depends on a certain virus. That is, as a *causal* theory or explanation of kin altruism: an attempt to say what it is that altruism among kindred depends upon. But a very different and indeed a contrary interpretation of the inclusive fitness theory is clearly possible.

Namely, it can be taken to be a *denial of the reality* of kin altruism, and a causal theory only of the *delusive appearances* of kin altruism which the surface of life presents in such abundance. It can be taken to mean that what superficially *appears* as kin altruism is, in reality, only the *selfishness of genes*, manifesting itself at the phenotypic level, between individual organisms. That is, when you *seem to* care for your son's or your sister's well-being, all that is really happening is that your genes are selfishly getting you, their puppet or tool, to work in *their* interest (which is simply the making of more future copies of themselves) *via* your son's or your sister's reproduction.

It is *this* interpretation of the theory, and this one alone (I need hardly say), which recommends itself to sociobiologists. Alexander, for example, writes that kin altruism, "by which the phenotype is used to reproduce the genes, may be described as phenotypically (or self-) sacrificing but genotypically selfish."[25] Dawkins writes that "a gene might be able to assist replicas of itself which are sitting in other bodies. If so, this would appear as individual altruism, but it would be brought about by gene selfishness."[26] It would be easy to multiply quotations to the same effect; but it can hardly be necessary.

Nor can it be doubted that it is this interpretation of inclusive fitness theory that sociobiologists *ought* to adopt. For the other interpretation of it—as a *causal* theory of kin altruism—leads (as we saw in Section III) to the most unsociobiological, and even the most un-Darwinian, of all possible outcomes. Namely, an enormous *over*estimation of the amount of altruism that there is in the world. Whatever sociobiologists mean to imply, and whatever Hamilton meant to imply, it is quite certain that any theory of theirs is not one which will err on *that* side.

Hamilton himself, in fact, gave his readers one very unmistakable hint that he meant to deny the reality of kin altruism, and to explain the appearances of it as due to the selfishness of genes. For, as the reader will see if he refers to the text to Note 3 above, he did not say that an animal does value, or will value, a neighbor's fitness against its own, according to its degree of relatedness to that neighbor. What he said was that the animal "will *seem to* value" its neighbor's fitness against its own according to (etc.).

What an amazing expression to choose! What an inexplicable one, unless Hamilton meant to advance a causal explanation, not of kin altruism, but only of the delusive *appearances* of kin altruism! The significance of the expression was certainly not lost on his disciples. Trivers for example, twenty years later, is still careful to insert, at critical places in his own writing, his master's "seem to."[27]

This was a clear enough indication that Hamilton intended the inclusive fitness theory as a denial of kin altruism, rather than as an explanation of it. But even apart from his intentions, his theory by its very nature leads inevitably, as it led in historical fact, to the sociobiologists' theory of "the selfish gene"; and hence, in particular, to the idea that what appears as kin altruism is really only the selfishness of genes.

A theory of evolution, in order to qualify as a Darwinian one at all, must resemble the original Darwinian theory at least to this extent: that it points out a class of causal agents whose interaction can reasonably be supposed to result in evolution. Before Hamilton, these causal agents were always supposed to be individual organisms; and the interaction among them, which would bring about evolution, was supposed to be the struggle among conspecifics to survive and reproduce. This old Darwinian picture may not have been true to life; but it was admirably definite and intelligible.

But all this definiteness and intelligibility vanishes the moment we start to think of organisms as always tending to maximize their *inclusive* fitness, rather than their own individual fitness. On the basis of that theory, what picture can we form of the process of evolution at all? In Hamiltonian Darwinism, what are the causal agents, and what is the interaction among them that is to "drive" evolution?

INDIVIDUAL ORGANISMS are now entirely out of the running. The individual is now no more than an insignificant speck in the vast cloud of its extended family or kin. This cloud is without any definite boundaries, in either time or space; without any real physical center (as distinct from the merely perspectival center which each organism supplies for itself); and without even the faintest suggestion of an internal structure, tendency, or cause, which might incline it to develop in one way rather than another.

Could these Hamiltonian kin clouds be *themselves* the causal agents whose interaction drives evolution? No. As causal agents for explaining anything, clouds—of mosquitoes, or of water droplets, say—are unpromising enough: they just don't interact with other things in a definite enough way. But a cloud of kin is far more unpromising still. After all, the mosquitoes or water droplets which make up a cloud have at least got to be at roughly the same place and time; but the members of a kin cloud are under no such necessity. In a kin cloud of elephants, or of humans, even two of the most closely related individuals—a father and daughter, say—can easily be separated by fifty years, or by one hundred miles, or by both. Any causal agents, which by their interaction are to make evolution "go," will need to be a great deal less diffuse than kin clouds are.

As causal agents, kin clouds are not only hopelessly diffuse: they are not even, in general, discrete. It "takes two to interact," but many kin clouds overlap so extensively as to prevent them being, in any intelligible sense, two at all. The kin cloud of even your cousins—the set of people who contribute to *his* inclusive fitness—*must* overlap extensively with your kin cloud. And as to *your brother's* kin cloud: how does it differ from yours *at all*, in such a way that the "two" things could meaningfully be supposed to interact?

All right: the causal agents whose interaction is to drive evolution cannot be the kin clouds which the Hamiltonian theory brings into view. Nor can they be the individual organisms, which that theory effectively pushes out of view. But then, what *are* the causes which bring about evolution, if the theory of inclusive fitness is true? If it really is a *Darwinian* theory of evolution, then this yawning vacancy in it must be filled by something.

Well, of course, by 1964, "there could be but one way" of filling it: *genes*. They were the only possible candidates for the position. They had some genuine qualifications for it, too. Causal agents they undoubtedly are; not at all diffuse ones, either, and even noted on the whole, ever since Mendel, for their high degree of discreteness. Interact with one another they certainly do, and that interaction certainly has a great deal to do, in detailed ways, with bringing evolution about. Whether their causal powers are sufficient, or even of the right kind, to make them the complete puppet masters in evolution which the sociobiologists think they are, may reasonably be doubted. But what of that? When a position *must* be filled, and there is only one possible candidate, the candidate can be sure of having his suitability for the position exaggerated, and in fact can "name his own salary."

This was the way in which Hamilton's theory of inclusive fitness gave birth to the selfish gene theory of the sociobiologists; and hence, in particular, to their belief that what seems to be kin altruism is really just gene selfishness. The process was one of the *a priori* exclusion of alternatives, rather than of empirical discovery. If the inclusive fitness theory is your *starting point*, as it was for the sociobiologists, then there just *are* no other causal agents which could drive evolution, and genes get the job for want of other candidates.

The inclusive fitness theory is therefore a denial of kin altruism, rather than a causal explanation of it. It is a new version of what I discussed in Essay 6: the immemorial selfish theory of human and animal nature. And it is not even, at that, a fundamentally new version of it.

Parental altruism was always, of course, the greatest stumbling block in the way of the selfish theory. But selfish theorists, at least as early as about 1700 A.D., had thought of a way round this difficulty. Namely, by saying that our "children are parts of ourselves, and in loving them we but love ourselves in them."[28] Which is clearly a simple prototype of the idea that what appears as parental altruism is really just a kind of selfishness.

It was effectively criticized, as long ago as 1726, by the Scotch philosopher Francis Hutcheson (from whom I have taken the quotation). But then, as I said in Essay 6, the selfish theory, however often refuted, never dies. It is probable, indeed, that this prototype of sociobiology is much older than the early eighteenth century. Men who were as flash as a rat with a gold tooth swarmed in the Greek Enlightenment of the fifth and fourth centuries B.C., and it would be surprising if none of them ever hit on this particular way of getting round the fact of parental altruism.

WE ARE NOW in a position to identify what the irresistible attraction is that the theory of inclusive fitness holds for every modern Darwinian mind.

If a man openly denies the reality of altruism, then, as well as incurring the deserved ridicule of people of common sense, he incurs the moral indignation of people of common decency; as Hobbes, Mandeville, and Machiavelli (among others) found out by experience. He deserves it, too. Now the Darwinian theory of evolution is a theory which logically impels whoever believes it to deny the existence of altruism. But for more than a hundred years (as we have seen), Darwinians all shrank from that denial: restrained, no doubt, partly by fear of the evil reputation of a Hobbes or Machiavelli, but also by their own decency.

But then came Hamilton's theory, and in the second half of the 1960s—those five fell years for Western civilization!—it brought into existence, in the way I have described, what is now called sociobiology. These Darwinians are not restrained either by common sense or common decency, and they do openly deny the existence of altruism. They thus announce themselves as the new Hobbists, Machiavels, or Man-devils (as someone called the followers of that once famous selfish theorist).

In consequence, sociobiologists have incurred at least a sizable fraction of the moral condemnation which they deserve. And they have failed (at least up till now) to carry with them, in their open denial of altruism, the great majority of their fellow-Darwinians, either professional or lay. Darwinians without exception nowadays, to be sure, swear by the theory of inclusive fitness. But very few let it lead them to say things of the kind which it leads the Hard Men of sociobiology to say. For example, "Scratch an 'altruist'" and watch a hypocrite bleed"; or "Nice Guys Finish Last"; or that conscience "tells us, not to avoid cheating, but how we can cheat socially without getting caught."[29]

But a denial of the reality of altruism which did *not* openly offend either common sense or decency: *that*, by contrast, would be exactly "what the doctor ordered" for all present-day Darwinians. It would give them what no Darwinians had ever had before: freedom to profess their Darwinism fully, without getting a bad name, and with a conscience that, if not quite unclouded, is not in revolt either. A combination "devoutly to be wished."

Now this combination is exactly what the theory of inclusive fitness *does* offer to Darwinians of the present day. For that theory is (as we have seen) a denial of the reality of kin altruism; but on the other hand it is not an overt one. It is covert, and in two ways. First, it is esoteric; since it is entirely in terms of genes, any knowledge of which *must* be esoteric. Second, it is indirect; since it directly ascribes selfishness not to people or other animals, but only to genes. Now, if *genes*, and only they, are accused of *selfishness*, what is there in that which could reasonably arouse the moral indignation of people of common decency, or even of one's own conscience?

This, then, is what constitutes the irresistible attractiveness of the inclusive fitness theory to every modern Darwinian mind: it allows you to deny the existence of altruism, even in its most conspicuous form, without giving unmistakable offense either to common decency or to your own self-respect. This had been a great desideratum ever since 1859 (as I have said): it was just that no one before Hamilton had been clever enough to find a way to do it.

The objections I made to the inclusive fitness theory in the previous section were made, of course, on the assumption that it is a theory of what causes kin altruism. In the present section, however, we have seen compelling reasons to think that it is not that, but is a denial of kin altruism, and a causal theory only of the *appearances* of kin altruism. If so, then my earlier objections were based on a mistaken assumption.

But they are very easily adapted so as to become good objections to the inclusive fitness theory as we now understand it to be. In fact all I need do is to insert the word "apparent" before the word "altruism" (and before "kin altruism," "parental altruism," etc.) in each of the objections which I made in Section III.

Thus, for example, where the objection previously was that sibling altruism is not nearly as common or strong as parental, it will now be that apparent sibling altruism is not nearly as common or strong as apparent parental altruism. Where the objection previously was that there is no altruism between genetically identical sister bacteria, it will now be that there is no apparent altruism between them. And so on.

Then every one of my objections will be found to be true still, and directed, *this* time, towards the right target. In other words, even if the selfish theorists of all ages are right, and there is no such thing as kin altruism, shared genes are just as bad an explanation of the *appearances* of kin altruism, as we earlier found them to be of the thing itself.

V

Yet almost every other line which inclusive fitness theorists write about altruism implies that I was right the *first* time: that is, that they *do* intend their theory as a causal explanation of kin altruism.

This is so evident everywhere that it would be ridiculous to assemble quotations to prove it. It will be sufficient if I draw attention again to some words of Dawkins quoted above (the text to

Note 2). He said that it had long been clear that the greater than average chance of a gene being shared with close relatives *"must be why* altruism by parents towards their young is so common." If this is not *causal* talk, then I cannot understand English. And it is perfectly representative of 50 percent of what inclusive fitness theorists write about kin altruism.

And yet these are the same authors who, in other parts of their books or articles, imply that altruism does not exist! They say, like Hamilton, that a mother *seems to* put a positive value on her baby's fitness against her own. Or they say, like Ghiselin, that so-called altruists are hypocrites. Or they refer, like Dawkins, to "altruism—something that does not exist in nature."[30]

What can we possibly make of this bewildering, and yet systematic, inconsistency? A proposition—the theory of inclusive fitness, or any other proposition—cannot be *both* a causal explanation of something and a denial of its existence. A causal theory of the tides cannot deny the reality of tides; nor can a causal explanation of polio myelitis imply that there is no such disease. A causal theory of x implies the existence of x, and cannot consistently deny it as well.

But that is not at all to say, alas, that people do not sometimes *confuse* a causal explanation of something with a denial of its existence. On the contrary they often do, as philosophers to their sorrow know. A man puts forward a causal theory to explain something, and another man thinks that he is denying the existence of that thing. Or a man *says* he is going to explain what causes something, and fails to notice that, in the course of his "explanation," he has implied that there is actually no such thing as that which he had undertaken to explain. Confusions of this kind are certainly common enough.

If you want an example of how easily people can confuse a causal explanation of something with a denial of its existence, then talk to a man in the street, or to an average physicist for that matter, about one of the secondary qualities: color, for example. Try to get him to be (a) clear and (b) consistent, about whether he is (1) putting forward a causal explanation of things having the colors they do, or (2) denying that things have any color. You will need to allow a good deal of time for this job. And it is a thousand to one that, however much time you allow, it will not be enough.

A third possibility therefore suggests itself. Perhaps the inclusive fitness theory is neither a theory of what causes kin altruism, as I took it to be in Section III; nor yet a denial of kin altruism, as I tried to show in Section IV that it is. Perhaps it is sometimes one and sometimes the other. Do inclusive fitness theorists just confusedly oscillate between thinking of shared genes as what causes kin altruism and thinking of shared genes as the reality which underlies the illusory appearances of kin altruism?

This hypothesis is somewhat disrespectful, I must admit, to inclusive fitness theorists. But then better scientists than they are certainly *have* often fallen into exactly this kind of confusion about color. So the disrespectfulness of the hypothesis can hardly be a decisive objection to it.

In fact this hypothesis has much to recommend it. For it would explain very well a certain feature of the literature of inclusive fitness, for which no other explanation suggests itself. The fact, namely, that that literature contains two violently inconsistent estimates of the amount of kin altruism that there is in the world.

An inclusive fitness theorist, if you accuse him of denying the reality of kin altruism, will almost certainly dismiss the accusation as an elementary misunderstanding of his theory. He will defend himself somewhat as follows. "*Of course* kin altruism is real, at the level of individual organisms. It is very common and strong, too: think, for example, of parental altruism in humans, of sibling altruism in hymenopteran workers, and so on. What Hamilton taught us was what it is that causes kin altruism: namely, the selfishness of genes."

This self-defense is certainly "full of whole wheat words";[31] no cynical suggestion of universal selfishness here. And, taking the inclusive fitness theorist at his word, we adopted his causal explanation of kin altruism in Section III above. It led us to a number of surprising results. For example, that there is twice as much sibling altruism between bacterial as between human sisters; that sibling altruism in our species is as common and strong as parental altruism; that *every* parent bird will sacrifice its own life in order to save three of its nestlings; and so on. And the combined result of all these discoveries was that there is in fact far more kin altruism in the world than anyone had ever supposed before the inclusive fitness theory came along. In fact it turned out that animal life is saturated with kin altruism: drips the stuff at every pore.

And yet, in the literature of the inclusive fitness theory, what do we actually find? Why, more often than not, the universality of "dog eat dog," of "dirty tricks," of the self-interested manipulation of offspring by their parents, of parents by their offspring, of siblings by each other, of strangers by everyone; of apparent altruism revealed as hypocrisy (even, no doubt, in those luckless hymenopteran workers who had previously been portrayed as paragons of kin altruism). There is no pretense, in *this* part of the literature, of admitting the reality of kin altruism and confining selfishness to the *gene* level. On the contrary, it is the Hobbesian war of all against all, openly installed (not for the first time) as the last word in Darwinian biology. There is not, it turns out, one atom of kin altruism in the world: it is an illusion.

In any discussion of the inclusive fitness theory with an adherent of it, the same extraordinary phenomenon of "Janus faces" will be met with. On one face of the theory, arising out of the idea that kin altruism is caused by shared genes, there is an extravagant exaggeration of the amount of kin altruism that exists; on the other, there is the idea that kin altruism is an *illusion*, the underlying reality of which is shared selfish genes. Any discussion of altruism with an inclusive fitness theorist is, in fact, exactly like dealing with a pair of air balloons connected by a tube, one balloon being the belief that kin altruism is an illusion, the other being the belief that kin altruism is *caused by* shared genes. If a critic puts pressure on the illusion balloon—perhaps by ridiculing the selfish theory of human nature—air is forced into the causal balloon. There is then an increased production of earnest causal explanations of *why* we love our children, *why* hymenopteran workers look after their sisters, etc., etc. Then, if the critic puts pressure on the causal balloon, perhaps about the weakness of sibling altruism compared with parental, or the absence of sibling altruism in bacteria—then the illusion balloon is forced to expand. There will now be an increased production of cynical scurrilities about parents manipulating their babies for their own advantage, and vice versa, and in general, about the Hobbesian bad times that are had by all.

In this way critical pressure, applied to the theory of inclusive fitness at one point, can always be easily absorbed at another point, and the theory as a whole is never endangered. A defender of the theory does need, it is true, a certain mental agility: an ability to make sudden and extreme "gestalt switches" (as the best authors in the philosophy of science now say), from a picture in which animal life is swimming in kin altruism, to one in which there is no kin altruism at all. But this ability, it has turned out, is by no means uncommon; and it is the only one which a defender of the inclusive fitness theory needs. Given that, his theory is stable under any criticism whatever.

My hypothesis—that inclusive fitness theorists are just confused about kin altruism, and oscillate between denying it and trying to explain it—has at least the merit, therefore, of explaining something otherwise improbable: the Janus-faced character of their theory. But it also has in its favor a historical fact which I point out in Essay 6: that selfish theorists have *always* oscillated between a version of their theory which is shocking but not true, and a version which is perhaps true, but certainly not shocking, or even interesting.

When an inclusive fitness theorist tells us that kin altruism does not exist, then that is shocking all right; but it is not true. When, on the other hand, he only tells us that kin altruism is caused by shared genes, then that happens not to be true (as we saw in Section III) but even if it were true, it would not be shocking, or even interesting. If kin altruism *is* caused by shared genes, that is well—it exists, anyway; if it is caused by something entirely different, well again. Who doubts that it is caused by *something*? Nor can its cause be of a very rare or elevated character, in view of the extreme commonness of kin altruism which (at least in its parental form) extends even to such low spirituality types as alligators. The fact that kin altruism has a cause does not prevent it from being sometimes an admirable thing, either. By *that* too severe rule, there would be nothing to admire anywhere; not even in, say, *The Selfish Gene*, which presumably had its causes like everything else in nature.

If inclusive fitness theorists do, as I believe, constantly oscillate between explaining and denying altruism, this must still further enhance the attractiveness of their theory to every Darwinian mind. For every such mind *needs* either an explanation or a denial of altruism. A theory which offers both of those things at once will therefore be doubly attractive.

And then, think how easy it is, and always has been, to convince many people of the selfish theory of human nature. It is quite pathetically easy. All it takes, as Joseph Butler pointed out nearly three centuries ago, is a certain coarseness of mind on the part of those to be convinced; though a little bad character on either part is certainly a help. You offer people two propositions: "No one can act voluntarily except in his own interests," and "No one can act voluntarily except from some interest of his own." The second is a trivial truth, while the first is an outlandish falsity. But what proportion of people can be relied on to notice *any* difference in meaning between the two? Experience shows very few. And a man will find it easier to mistake the false proposition for the evidently true one, the more willing he is to believe that everyone is as bad as himself, or to belittle the human species in general. (Darwinians call the latter "bridging the gap between man and the animals.")

It is even easier nowadays to convince people that, even within families, there is nothing but selfishness. All you need to do is tell them that "what appears as kin altruism is really gene selfishness." If "appears as" means here "seems to be but is not," then the statement is a denial of the reality of kin altruism. But "appears as" can also mean "result in," or "has as one of its effects": as when we say, for example, that the moon's gravitational attraction appears as tides in the ocean. If this is what "appears as" means here, then the statement is a theory of what causes kin altruism. But what proportion of people can be relied on to notice this ambiguity of the phrase "appears as," or to notice the result of it: that the given statement can equally well be a denial of kin altruism, or a causal explanation of it? Again, very few.

Well, how could inclusive fitness theorists *not* oscillate between those two things? If they were to adhere consistently to the causal version of their theory, the result would be (as we saw in Section III) far *more* kin altruism than actually exists, and Darwinism's problem of altruism would actually be far *worse* than it was before Hamilton supposedly solved it. If they were to adhere consistently to the denial of the reality of kin altruism, that would indeed solve the problem of altruism with supreme *éclat*, but would be considered by everyone of common sense to be a *reductio ad absurdum* of their theory, rather than a scientific discovery. So what can inclusive fitness theorists do, except what they do do? That is, publish hundreds of articles every year, in which kin altruism is both denied *and* causally explained in terms of shared genes. These two things may be logically inconsistent with one another. But what of that? It's a "successful research program," isn't it?

At any rate *some* explanation is required of the Janus faces of the inclusive fitness theory: on one side an immense exaggeration of the extent of kin altruism, on the other a denial that there is any at all. My hypothesis is: mere confusion in the minds of inclusive fitness theorists. This is, at least, a better suggestion than the only other hypothesis I have been able to think of: that inclusive fitness theorists deliberately try to *deceive* their readers, by passing off a denial of kin altruism as a causal explanation of it. This is an eminently sociobiological hypothesis, of course; but like all such things, it has nothing to recommend it. Confusion is always more likely than elaborate cunning and "dirty tricks."

So if (for example) you cannot work out by reading *The Selfish Gene*—as you cannot work out—whether the author is denying the existence of kin altruism or offering a causal explanation of it, then easily the likeliest reason is that the author's thoughts on that subject are in exactly the same state of incoherence as his book is. It is a rule with very few exceptions that the book *is* the man; and even, the man when intellectually at his best.

Besides, if that book had said, clearly and consistently, either that kin altruism does not exist, or that it does, how much of its piquancy, and its sales, would have been lost! Its inconsistency on this fundamental point, while no doubt faithfully reflecting the author's mind, was one of the very things which kept its readers interested and guessing. A source of interest to the readers, it was a source of income to the writer, and consistency would have cost him money.

Essay 9
A New Religion

. . . that sacred particle [the seed].
—Paley, *Natural Theology*

I

DOLPHINS and some other animals have lately turned out to be more intelligent than was formerly thought, and present-day computers are capable of some amazing things. Still, if the question is asked, what are the most intelligent and all-round capable things on earth, the answer is obvious: human beings. Everyone knows this, except certain religious people. A person is certainly a believer in some religion if he thinks, for example, that there are on earth millions of invisible and immortal non-human beings which are far more intelligent and capable than we are.

But that is exactly what sociobiologists do think, about genes. Sociobiology, then, is a religion: one which has genes as its gods.

Yet this conclusion seems incredible. Was not religion banished from biological science a long time ago? Why, yes. And is not sociobiology a part of biological science (even if a very new part, and a controversial one)? No. Sociobiologists really are committed to genes being gods, as I will show in a moment.

But first consider the following. We would all say, because we all know it to be true, that calculating machines, automobiles, screwdrivers and the like, are just tools or devices which are designed, made, and manipulated by human beings for their own ends. Now, you cannot say this without implying that human beings are more intelligent and capable than calculators, automobiles, screwdrivers, etc. For if we designed and made something as intelligent and capable as ourselves, or more so, it would be precisely *not* just a tool which we could manipulate for our own ends: it would have ends of its own, and be at least as good at achieving those ends, too, as we are at achieving ours. Similarly, suppose someone says that human beings and all other organisms are just tools or devices designed, made, and manipulated by so-and-sos for *their* own ends. Then he implies that so-and-sos are more intelligent and capable than human beings.

With that in mind, consider the following representative statements made by leading sociobiologists. Richard Dawkins, easily the best-known spokesman for this movement, writes that "we are . . . robot-vehicles blindly programed to preserve the selfish molecules known as genes,"[1] and again that we are "manipulated to ensure the survival of [our] genes."[2] The same writer also says that "the fundamental truth [is] that an organism is a tool of DNA."[3] (That is, of the DNA molecules which are the organism's genes.) Again, Dawkins says that "living organisms exist for the benefit of DNA."[4] Similarly, E. O. Wilson, an equal or higher sociobiological authority, says that "the individual organism is only the vehicle [of genes], part of an elaborate device to preserve and spread them . . . . The organism is only DNA's way of making more DNA."[5]

I will mention in a moment some other passages in which sociobiologists imply that genes are beings of more than human intelligence and power, but that implication should be clear enough already from the passages just quoted. According to the Christian religion, human beings and all other created things exist for the greater glory of God; according to sociobiology, human beings and all other living things exist for the benefit of their genes. The expression "*their* genes" is probably not perfectly orthodox, from the strict sociobiological point of view, being rather too apt to suggest that genes are part of *our* equipment, whereas (according to sociobiology) we are part of theirs. All the same, the religious implication is unmistakable: that there exist, in us and around us, beings to whom we stand in the same humble relation as calculators, cars, and screwdrivers stand in to us.

It must be admitted that sociobiologists sometimes say other things which are inconsistent with statements like the ones I have just quoted. Dawkins, for example, sometimes protests that he does not at all believe that genes are "conscious, purposeful agents."[6] But these disclaimers are in vain. *Of course* genes are not conscious purposeful agents: everyone will agree with that. Where sociobiologists differ from other people is just that they also say, over and over again, things which imply that genes *are* conscious purposeful agents; and agents, at that, of so much intelligence and power that human beings are merely among the tools they make and use.

It is in Richard Dawkins' book *The Extended Phenotype* that the apotheosis of genes has been carried furthest. Manipulation is the central idea of this book (as the author himself acknowledges),[7] and more specifically, manipulation by genes. Genes are here represented as manipulating, in their own interest, not only the bodies and behavior of the organisms in which they sit, but just about everything under the sun.

Genes manipulate external objects. For example, spider genes (not spiders) manipulate webs, termite genes (not termites) manipulate mud to make their mounds, beaver genes (not beavers) manipulate logs and water to make a dam, and so on.[8] Action at a distance, something which is usually considered to be difficult or impossible, is no trouble at all to genes.[9] No job is too big for them, either: beaver genes can easily build a lake miles wide.[10] Genes also manipulate the behavior of other organisms, and the victims of their manipulation need not at all be of the same species as the organisms which carry the manipulating genes.

For example, a certain kind of cuckoo deposits its egg among the eggs laid by a reed warbler. Once the eggs hatch, the excep-

tionally loud begging cry of the young cuckoo and its exceptionally colorful "gape" induce the parent reed warblers to give it
more food than they give to their own young. According to
Dawkins, this is a case of the genes of the cuckoo parents
manipulating the behavior of reed warbler parents, to the advantage of the former and the disadvantage of the latter.[11]

Now, think what this kind of description commits the user of
it to. Just as maternity implies parenthood but not conversely, so
manipulation implies causal influence but not conversely. The
moon causally influences the tides, but it cannot manipulate
them. Even if causal influence results in some advantage to the
influencing agent, that is still not enough to constitute manipulation. If you and I are competing to catch the greater number of
fish from our boat, and I by accident knock you overboard, then
I influence your behavior but do not manipulate it, even though
your mishap improves my chances of winning the competition.
To constitute manipulation, there must be the element of *intended or purposeful* causal influence.

Most biologists would see, in a case like nest parasitism,
nothing more than an extremely complex example of causal influence. They might ascribe to the genes of the cuckoo exactly the
same causal influence as sociobiologists do. What distinguishes
the sociobiologist's description of the case is his insistence that
those genes are *manipulating* the reed warblers' behavior for
their own benefit. Well, cuckoos do benefit, and reed warblers
lose, by nest parasitism. But, as we just saw in the boat case,
causal influence plus resulting advantage are not enough to constitute manipulation. The causal influence must also be purposeful or intended. But is that condition satisfied in this case?

If the nest parasitism of cuckoos *is* a case of manipulation, it is
certainly a staggeringly clever one: far too clever for cuckoos, in
particular, to be capable of. Can a cuckoo *have* a purpose as
complicated as that of getting a reed warbler to feed a cuckoo
nestling better than it feeds its own young? That must be extremely doubtful. Still, let us suppose that a cuckoo is clever
enough for that. He would need to be cleverer still, to be able to

think up a way of *achieving* this purpose. In particular, could he think up a way of achieving it which did not involve any cuckoo's ever going even within a mile of a reed warbler? No: there is no one who will credit cuckoos with so great an intellectual feat. Yet even if a cuckoo could manage that part too, the hardest job would still lie ahead of him. For he would need, not just to *have* this brilliant idea, but to be able to implement it. But how is a cuckoo to do whatever engineering is required? He has no hope. Manipulative ability of any kind is not highly developed in birds, and cuckoos are distinctly below the bird average in this respect. After all, hardly any of them can even build a nest.

But the feat of manipulation in question would not only be too hard for cuckoos: it would be too hard for us.

Suppose that nest parasitism has not yet evolved among birds, and that young cuckoos have not yet acquired their special adaptations for it. Cuckoos (we will suppose) raise their own young, but are extremely slapdash parents. In these circumstances, we might become anxious about the survival of cuckoos, and decide to take steps to improve their reproductive performance.

Now, would you or I be clever enough to think of nest parasitism as a means to this end? I know I never would; but perhaps you would. But would even you be able to think of a way of getting the host birds not only to feed the young cuckoos, but to feed them better than their own offspring? A way, at that, which does not require any human cuckoo helper ever to go near a member of the host species? With all due respect to human intelligence, this seems hardly possible. Still, let us suppose that we did think up such a way, and that in particular we came up with the brilliant idea of endowing young cuckoos with exceptional voice and gape. Even then, the hardest part of the job would still remain: that is, to *implement* this idea. Well, human beings are as preeminent on earth for engineering ability as they are for intelligence, but we could not do this. We cannot build young cuckoos, or breed them, to precise specifications. And no genetic engineer could as yet undertake this particular task with rational confidence of success.

It would, then, be a feat of manipulation, not only far beyond cuckoo capabilities, buy beyond present human capabilities, to prevail on reed warblers, without having to go near them, to feed cuckoo young at the expense of their own young. Yet this feat is one which, if Dawkins is right, cuckoo *genes* first performed long ago, and have practiced ever since without the smallest difficulty. The implication could hardly be plainer: cuckoo genes are more intelligent and capable than human beings. The same presumably holds *a fortiori* for human genes.

The only way in which sociobiologists could avoid this implication would be if they used the word "manipulation" when they ascribe manipulation to genes, in some sense which does *not* imply purpose as its ordinary sense does. But they do not do any such thing. Dawkins, for example, makes no distinction whatever between the manipulation which he ascribes to genes, and the ordinary sense of the word, in which he says (as we all say) that pigeons manipulate twigs and other nest materials, beavers manipulate logs to build a dam,[12] and so on.

Gods, in addition to being thought of as more intelligent and powerful than we are, are always thought of as being immortal. It was therefore to be expected that sociobiologists would wish to ascribe this attribute, too, to genes. Here is a passage from Richard Dawkins on this subject. The "gene . . . does not grow senile; it is no more likely to die when it is a million years old than when it is only a hundred. It leaps from body to body down the generations, manipulating body after body in its own way and for its own ends, abandoning a succession of mortal bodies before they sink in senility and death. The genes are the immortals . . . ."[13]

II

Most people would *like* some religion to be true. This may seem strange, when you consider that every religion is and must be more or less terrifying. But then, there are various things which can outweigh terror. One of them is depression, and if religion is terrifying atheism is depressing. It is an intensely depressing thought that the brightest and best things the universe has to show are certain members of *our* species.

The trouble is, though, that every religion (or at any rate every one I know of) is incomprehensible when it is not obviously false. Of course, something which is incomprehensible to us might nevertheless be true, and religious people often remind the non-religious of this fact. But, though it is a fact, it is no help, because there are always many *competing* incomprehensibilities, from religious and other sources, vying for our acceptance. Tertullian said that he believed the Christian religion because of its absurdity. But alas, every other religion possesses the same claim on our belief (if absurdity really is a claim on our belief).

Sociobiology is not incomprehensible, but it is one of the religions which are obviously false. The only part of it that is true is the doctrine that genes are invisible. But this is not something peculiar to sociobiology. Everyone agrees that genes are invisible, at least to the naked eye and to old-fashioned microscopes. Given present-day microscopy, however, any invisibility which genes can still be said to possess is invisibility of no very deep or interesting kind. (It is not at all like the invisibility of *numbers*, for example.)

Sociobiologists have consciously and avowedly revived the doctrine of the "immortality of the germ-plasm" (or of the "germ-line"), which August Weismann first published about a hundred years ago. (Dawkins, indeed correctly describes his own overall position as "extreme Weismannism.")[14] But that famous doctrine is, and always was, true only in a highly special and indeed idiosyncratic sense.

The extinction of a species (that is, its last member dying) is a common enough occurrence in evolutionary history; and every time it happens, every gene-line peculiar to that species comes to an end too. When you die, that is also the end of every gene-line which any cell of yours had been carrying on until then. If a man has no sons, then many of his genes—namely, at least all the ones on his Y chromosome—are not transmitted any further. If you have no children at all, you are not a node on *any* gene-line which extends into the future. Not a very robust kind of immortality, this! In fact I would be thinking seriously, if I were a gene, of bringing a suit for misleading advertising against Richard Dawkins and the Weismann estate.

The "immortality" of genes or of gene-lines, then, if not an obvious falsity, is an exceedingly misleading expression. (Dawkins himself, having said in the passage quoted earlier, that "the genes are the immortals," was prudent enough to add at once that they are not *really*.)[15] The grain of truth in the doctrine of gene immortality is that genes, unlike most organisms, do not have a natural life cycle ending in death. But this truth was never anything to write home about, since genes are not organisms, are not alive, and (as far as I know) have never even been thought to be so. The world "immortal" means "alive and not subject to death," and it therefore can be properly applied only to something living. To apply it to things which are not alive, such as DNA molecules, can only serve to bewilder oneself and other people; for example, by encouraging misplaced feelings of awe towards genes.

Genes thus lack both the durability and the life which would be needed to justify calling them "the immortals." But even if they had been alive, and much more durable than in fact they are, this would do little to make the gene *religion* credible. In the godhood stakes, the superior durability of a life form, if not accompanied by superior intelligence and capability, does not count for much. If it did, carp would be more godlike than horses, and it would be a close run thing between human beings and elephants. But it is not so.

The main reason, however, for thinking that sociobiology is false is the simple one I gave at the beginning: that it is obvious that human beings are the most intelligent and capable things on earth. But genes are not human. Therefore (etc.).

Genes are so far from being the winners in the intelligence capability stakes, that they are not even starters. They are just molecules of DNA, after all, and DNA molecules have exactly as much intelligence and purpose as (say) H_2O or $NaCl$ molecules:

namely none. They differ from almost all other molecules in having a strong tendency to produce copies of themselves. But then, ever since we first notice that the offspring of human beings are human, that the offspring of mice are mice, etc., we might have known that there must be *some* physical mechanism by which parental characters are transmitted to offspring. It need not have been gene replication, but it did have to be some sort of machinery for producing the same thing again. Of the details of this machinery, a great deal is now known. But this part of science has not brought any gods to light. On the contrary, like every other part of science, it has only served to drive those elusive beings still further into the shade.

Genes, even if they were alive, and did possess intelligence and purpose, would still be hopelessly miscast in the role of the world's greatest manipulators. Manipulation requires not only influence and intention to influence, but *means*. Yet what means of manipulation have genes got? They are *sans* limbs, *sans* organs, *sans* tissues, *sans* nerves, *sans* brains . . . *sans* everything. If they were capable, under these crushing handicaps, of any sort of manipulation, they really would have a good deal of the god about them.

III

It is logically possible (as should go without saying) that the sociobiologists are right and I am wrong. There is nothing objectionable *a priori*, or philosophically, about the proposition that genes are the most intelligent and capable things on earth. It is a question of fact, and nothing else, whether they are or not.

If they are, it will be an immense historical irony. Religion, which was driven out of biology by nineteenth-century Darwinism, will have been put back by—of all people—the extremists of neo-Darwinism.

This seems hardly conceivable, because for more than 2,000 years science has been at war with religion. Yet if the sociobiologists are right, science has actually now brought us what the human heart has always yearned for but never before achieved: knowledge of beings which, in virtue of their immense superiority to ourselves, are proper objects of our reverence and worship.

Essay 10
Paley's Revenge, or Purpose Regained

... the organism chooses its own effective environment from a broad spectrum of possibilities. That choice is *precisely calculated* to enhance the reproductive prospects of the underlying genes. The succession of somatic machinery and selected niches are *tools and tactics for the strategy* of genes.

—G. C. Williams, *Adaptation and Natural Selection,* 1966

I

IN ESSAYS 7 and 9 I have tried to show that the present-day demonology or religion of genes is not true. My object in the present essay is to explain how it came about.

The *last* part of this historical explanation has already been given, in Essay 7. This theory of inclusive fitness, I there pointed out, solved (at least to the satisfaction of Darwinians) the previously insoluble problem of altruism. It did so (as we saw) by resolving the *appearances* of kin altruism into the reality of gene selfishness. But this was only the last stride of the long journey which terminated in the new religion. It was the keystone which is placed at the top of an arch when, but only when, all the other stones are already in place. Hamilton contributed the idea that genes are selfish; but the other main element of the new religion—that organisms are merely tools manipulated by agents of more than human intelligence and power, their genes—was already in place *before* the theory of inclusive fitness was published in 1964.

Anyone can easily satisfy himself of this fact if, after having read Dawkins' *The Selfish Gene*, he reads *Adaptation and Natural Selection*, by George C. Williams.[1] This book was published ten years before Dawkins', in 1966, and the writing of it, we learn from the preface, "dates largely from the summer of 1963."[2] That is, before Hamilton's theory had arrived on the scene. Williams immediately and gratefully incorporated the inclusive fitness theory into his book,[3] and he clearly recognized the "keystone" position which it occupies in the intellectual structure of the book.[4] And that incorporation having been made, *Adaptation and Natural Selection* contained *all* the intellectual elements of the new religion of selfish genes.

In fact the only thing which was then lacking, in order to bring *The Selfish Gene* into existence, was the talent for vivid expression and popularization which Richard Dawkins possesses to a marked degree. Williams lacks this talent as abundantly as Dawkins has it. He is in no way a bad writer, but it is impossible to mistake him for a lively one. Dawkins properly acknowledged, in *The Selfish Gene*, the heavy intellectual debt which that book owes to Williams' books.[5] Many other authors have likewise paid deserved tribute to the influence of *Adaptation and Natural Selection,*[6] and it is, beyond question, the book which did far more than any other to inaugurate the new religion.

Its subject, however, is not altruism. It is something which lies equally close to the heart of Darwinism, and is far more widespread and prominent in organisms than altruism is: namely, adaptation. Organisms differ from inanimate objects in being, in countless ways, adapted or adjusted or fitted to the circumstances which surround them. Every one of their organs, structures, processes, phases, has a function or purpose: something that it is *for*. It is in order to explain *this* great fact of life, and to explain it along the most severely Darwinian lines, that *Adaptation and Natural Selection* is written.

II

I happened one day to be looking at a certain leaf of a tree, on which a bird dropping had fallen and dried: a sufficiently uninteresting object. But all at once the "bird dropping" turned into a spider. It was exactly as might happen in a dream, and as does happen in nursery stories: "the toad turned into a prince," etc.

Afterwards, of course, I found out that there are many species of spiders, and of insects too, which mimic bird droppings. But at the time I had never heard of this particular form of mimicry. After a few seconds, during which my initial stupefaction wore off, what I *thought* was this. "What an incredibly brilliant idea! Who's afraid of a bird dropping? Who wants to eat a bird dropping, or otherwise disturb it? No one: or no one, at least, who is either a potential prey of spiders, or a potential predator of spiders. Mimicking a bird dropping, as a way in which a spider can deceive both its prey and its predators, is so clever that I would never have thought of it, if you had given me a million years." And I am sure that anyone else, in the same circumstances, would have thought exactly the same.

But of course every adaptation, of any organism whatever, makes on us exactly this same kind of intellectual impression, once we realize that the thing in question *is* an adaptation. We then see that the thing fulfils a function, and does so in a *clever* and *well executed* way. At least, if we do not literally see these things, then our experience is so exactly *like* seeing them, that it is introspectively indistinguishable from it.

David Hume possessed the most penetrating critical intelligence of any modern philosopher, and he was the most unsparing of all critics of religion, especially in his posthumously published *Dialogues Concerning Natural Religion* (1779). The principal target of this book was "the design argument" for the existence of God: that is, the argument for an intelligent, powerful, and benevolent creator, from particular contingent features of the world, and especially from the adaptations of organisms. But Hume was far too candid a man not to acknowledge that when we come face to face with any concrete adaptation, we *cannot help* thinking that a purpose has here been achieved in a way which only high intelligence could have thought of, and only high engineering ability could have put into practice. He writes: "Consider, anatomize the eye; survey its structure and contrivance; and tell me, from your own feeling, whether the idea of a contriver *does not immediately flow in upon you with a force like that of sensation.*"[7] And of course, the only possible candid answer is that it does.

But now, in the case of the bird dropping spider for example, who or what could this contriver be?

Certainly not any *human* being, anyway. We have no interest whatever in helping spiders to elude predators or capture prey. Even if we did have, we would almost certainly never have been clever enough to think of mimicking bird droppings as a means of accomplishing those ends. But even if we had been clever enough to think of that, we would quite certainly not have been able to *do* it. We have not the faintest idea of how to make a spider which does not look like a bird dropping into one which does.

Could the contriver have been the spider itself, or its spider parent—or other ancestors? Well, it is not too much to suppose that spiders want to elude predators and to capture prey: so there is no difficulty in ascribing these *purposes* to them. But what of the intelligence required, to think of mimicking bird droppings as a means of achieving that purpose, and of the engineering ability required, even if a spider could *have* that thought, to implement the thought? Both of these things seem to be immeasurably beyond the abilities of spiders or their ancestors. After all, spiders and almost all other animals, outside the fixed range of the adaptations which they do exhibit, seem to be of an almost inconceivable degree of stupidity or incapacity. So how could spiders or ancestors of them, supposing that they had once lacked the adaptation of mimicking bird droppings, ever have acquired it?

But even if we managed to believe that some present-day species of spiders or their ancestors did think up the adaptation of mimicking bird droppings, and gradually accomplished this feat of engineering, far worse difficulties would still lie ahead of us. For countless adaptations of *plants*—for example, attracting insects to fertilize them, for dispersal of their seeds, and so on— are just as staggeringly brilliant as any adaptations of animals. And even if we can bring ourselves to believe that plants have purposes, it is impossible to believe that they have either high intelligence or great engineering ability.

All right: but if the intelligence and the ability displayed in the bird dropping mimicry of certain spiders were not those of human agents, nor yet those of the spiders themselves or their ancestors, whose *were* they? The intelligence displayed in this case probably exceeds human intelligence, and certainly exceeds the intelligence of spiders or their ancestors; and the engineering ability displayed enormously exceeds human ability, which in turn far exceeds the engineering ability of any other animals. What *can* we conclude, then, except that the bird dropping mimicry is the accomplishment of a purpose of some agent far more powerful and *intelligent* than any human? That is, that this adaptation, and every other, is due to divine purpose, intelligence, and power.

III

Such was, in essence, the famous old "design argument" for the existence of God, which received its classic formulation in William Paley's *Natural Theology* (1802). But of course Paley did not invent the argument. For centuries before he wrote, it had been carrying conviction to almost every rational and educated mind.

It continued to do so for another fifty years *after* Paley wrote. This is a historical fact which deserves to be known and reflected upon, yet it has been almost completely forgotten. Far from having suffered a fatal blow at Hume's hands in 1779, the design argument entered the period of its greatest flourishing only between 1800 and 1850. In 1829, for example, the Earl of Bridgewater provided a large sum in his will for a series of books to be written by the ablest authors, which would argue, not from revelation or from authority but rationally, for "the Power, Wisdom, and Goodness of God, as manifested in the Creation."[8]

The "Bridgewater treatises" duly came to be published, and they *were* written by the best authors. In retrospect, one in particular stands out. This was *The Hand* (1833) by Sir Charles Bell: the greatest of all British physiologists after Harvey. Yes, that's right: a whole book on the human hand, as evidence of the existence, intelligence, power and benevolence of God, only twenty-six years before *The Origin of Species* appeared! And it is—even if no one in the whole world now cares to know the fact—a very good book indeed.

By 1850, however, there were flickers of evolutionary lightning all around the horizon. Then, in 1859, the storm broke. Every reader of *The Origin of Species* could see at once that Darwin had put forward an explanation of adaptation which was new, simple, and ingenious. According to his theory, adaptation is not a result of divine, or of any, purpose, intelligence, or engineering skill. It is an effect of altogether *blind* forces: namely, the pressure of population, variation, and the resulting struggle for life among unequally endowed competitors.

The ancestors of the present-day bird dropping spiders did not possess that particular form of mimicry. Then, a variation cropped up among a few of them, which consisted in a faint resemblance to a bird dropping. This attribute would tend to make the spiders which possessed it better at capturing prey, and at eluding predators, than those which lacked it. They would therefore tend to live longer and leave more offspring, some of whom would inherit this useful attribute, than their less favored competitors. If other things were equal between those spiders which possessed this resemblance and those which did not, the former would, barring accidents, become progressively more common in the population, the latter less so. Ultimately, natural selection would ensure that *all* spiders of the species possess the bird dropping mimicry. Then, in the same way, the individuals which resembled a bird dropping a little more *closely* than their competitors did would enjoy a reproductive advantage over those whose resemblance to a bird dropping was less close. Natural selection would thus bring about a progressively more and more accurate mimicry of bird droppings. In time, the whole species would come to consist of spiders whose resemblance to a bird dropping was so uncannily accurate that *to us* it would look for all the world as though, for the purpose of enabling certain spiders to survive and reproduce, there had been an exercise of more than human intelligence and skill.

Such was, in essence, the Darwinian explanation of adaptation. In addition to its intrinsic merits, it had the advantages, over the Paleyan or theistic explanation, of being completely down to earth, and of explaining many other things *beside* adaptation. After all Darwin, in the *Origin*, had not been trying to explain adaptation: he had been trying to explain the origin of species! And yet, as Williams observes, the natural selection theory is actually a *better* explanation of the preservation and accumulation of adaptations, than it is of the origin of species.[9]

This explanation of adaptation, because of its obvious merits, immediately carried the day. The prestige of the theistic explanation, and of Paley with it, fell at once into a steep and apparently irreversible decline. "Natural theology," "religion independent of revelation"—the great defensive outwork which eighteenth-century Christians had built in order to set a limit to the advance of atheism—found that its principal support had been removed. What Hume had tried to do in 1779, but failed, Darwin succeeded in doing in 1859, without even trying.

By 1960 the reputation of the design argument, and of Paley, had been in free fall for a hundred years, and everyone with the smallest tincture of education "knew" by then that the theistic argument from design is beneath contempt, and that Paley was a fool or hypocrite or both. Only someone who has tried in recent decades, as I have, to convince silly undergraduates of the merits of Paley's classic book, can appreciate the absolute impossibility of that task. Paley was a Christian and (worse) a clergyman, he was on the opposite side to Darwin, and anyway (most important of all) he *lost*: that is "all they know, and all they need to know" of his matter.

But that attitude is really just part of the silliness of such people. And, as it happens, it has met with the punishment which it all along deserved. For in the last thirty years, Paley has had his revenge on Darwinism for more than a century of undeserved contempt. The explanation of adaptation by reference to the purposes of intelligent and powerful agents has come back into its own. And its reinstatement has turned out to require only some comparatively minor changes to the theology involved.

It is important to realize (and pleasant to record) that the vulgar contempt for the design argument was never shared by Darwin, or by any intelligent Darwinians who belong to what might be called "the pure strain" of intellectual descent from him. Well, this fact might have been anticipated. In any game, the formidable players are the best judges as to which of their opponents are formidable, and which are not.

When he was an undergraduate at Cambridge, Darwin was required to study Paley's *Evidences of Christianity* (1794). He tells us in his autobiography that "the logic of this book and, as I may add, of his 'Natural Theology,' gave me as much delight as did Euclid." Again: "I do not think I hardly ever admired a book more than Paley's 'Natural Theology.' I could almost formerly have said it by heart."[10]

Richard Dawkins, likewise, is full of a proper respect for Paley's explanation of adaptation. He even thinks so well of it that he cannot, he tells us, "imagine anyone being an atheist at any time before 1859."[11] He is scornful of those philosophers who claim that it was *Hume* who disposed of the design argument, by the suggestion in his *Dialogues* that adaptation needs no special kind of explanation. Dawkins' scorn is entirely justified. Anyone who recognizes (for example) the mimicry of a bird dropping as an adaptation of spiders, and yet says that it requires no different kind of explanation from that required by (say) the presence of iron in certain rocks, thereby stamps himself as belonging to the class of *uncandid* reasoners. He is not in earnest himself; and he therefore cannot reasonably complain if other people, who are in earnest, decline to take him seriously.

It is not in the least surprising that Dawkins should feel a profound intellectual sympathy with Paley's great book. It would be astounding if the opposite were the case. For he is a theist himself, as I have pointed out in Essay 7 and 9. He *agrees with* Paley that the adaptations of organisms are due to the purposive agency (more specifically, the selfish and manipulative agency) of beings far more intelligent and powerful than humans or any other organisms.

Dawkins has some disagreements with Paley, of course; but this really is a matter of course. When did two theists ever agree on *all* points? For example, Paley believed in the benevolence of God: see his Chapter XXCI, "Of the Goodness of the Deity." Dawkins, on the other hand, as we saw in Essay 7, ascribes to the gods of his religion a ruthlessly selfish character.

Then, Paley, being a Christian, believed (see his Chapter XXV) in "The Unity of God"; whereas Dawkins is a polytheist, as any adherent of the gene religion must be. But after all, the precise *number* of the gods is a comparatively minor point. Let it be one, or three, or 30,000 (as Hesiod computed), or a number rather larger than that (as gene religionists believe). The great, the fundamental point of religion is, rather, and always has been, the *existence* of purposive beings of more than human intelligence and power. And as to that, Dawkins and Paley are in agreement.

In any case, the monotheism of the Christian religion has never been anything to write home about. Or rather, to tell the truth, its "triune" God has been its Achilles' heel all along, and a perpetual source of scandal, not only to Jewish or Moslem minds, but to the countless sensible Christians who cannot help thinking that three and one are different numbers. And on the other hand, what assurance have we, that the *gene* religion will not be, ten years from now, a great deal less polytheistic than it is at present? The present structureless and disorderly democracy of selfish genes hardly looks like the last word of the Darwinian revelation. Genes are certainly older than any organisms, and according to Williams, "the DNA molecule has all the appearances of an evolved adaptation."[12] One would therefore expect genes to have evolved, long before now, some *general* structure much more comprehensive than the petty local "bossisms" or dominances of some genes over their alleles: something analogous to the vast taxonomic system which their organic vehicles belong to. There may well be, then, just around the next intellectual corner, an orderly hierarchy of gene "principalities and powers," or even the absolute monarchy of *one* Gene. (Odd if it turned out to be a triune one.)

Dawkins' enthusiasm for Paley, and for putting purpose back into the explanation of adaptation, great as it is, is thrown completely into the shade by that of his mentor, Williams. In *Adaptation and Natural Selection*, there are literally hundreds of sentences, and sentences which contain the very essence of the book too, which it would puzzle any reader to say whether they are more reminiscent of *The Selfish Gene* or of Paley's *Natural Theology*. And the reason is (as I have indicated) that in ascribing adaptation to divine purposes, those two books are one; while *The Selfish Gene* owes most of its intellectual substance to Williams' book.

Williams has a pet aversion, which he is always returning to castigate. This is, the failure of many of his fellow Darwinians to distinguish between the *function* of an organ, structure, process or whatever, and mere *effects* which it may have. A stock example (though not one Williams uses) concerns the heart. A heart, whenever and only when it circulates blood, also makes a certain sound. But the *function* of the heart's beating is to circulate blood; not to make a sound, which is merely an effect of the heart's beating.

A function or adaptation is something which "is produced by

design, and not by happenstance."[13] In particular, Williams insists, it is not enough to prove that something is an adaptation,
that it is *beneficial* to the organisms which possess it. "The
demonstration of benefit is neither necessary nor sufficient in the
demonstration of function . . . . It is both necessary and sufficient
to show that the process is *designed to* serve the function."[14]
"[T]he demonstration of effects, good or bad, proves nothing. To
prove adaptation one must demonstrate a *functional design*."[15]
Could Paley himself have said fairer than all this?

Here are some more passages which are fully representative of
Williams' book, in that they point equally to the Paleyan explanation of adaptation by super-human purposeful agents, and
to the present-day identification of those agents with genes.

"[E]very adaptation is *calculated* to maximize the reproductive
success of the individual, relative to other individuals . . ."[16] An
adaptation is "a mechanism *designed to* promote the success of
the individual organism, as measured by the extent to which it
contributes genes to later generations of the population of which
it is a member."[17] "Each part of the animal is *organized for* some
function tributary to *the ultimate goal* of the survival of its own
genes."[18]

Williams once or twice writes as though the purposes which
bring about adaptation are purposes of individual organisms. For
example, "the goal *of the fox* is to contribute as heavily as possible to the next generation of a fox population."[19] But this is no
more than an occasional *façon de parler*. The book as a whole
leaves us in no doubt that it is not organisms, but genes, which
design or calculate or organize adaptations. Foxes, seals, etc., are
not designers: they are designed. "[S]eals were *designed to*
reproduce themselves, not their species."[20] "[T]he *real goal* of
development is the same as that of all other adaptations, the
continuance of the dependent germ plasm."[21] "[T]he organism
chooses its own effective environment from a broad spectrum of
possibilities. That choice is *precisely calculated* to enhance the
reproductive prospects of the underlying genes. The succession of
somatic machinery and selected niches are *tools and tactics for*

the strategy of genes."[22] Could Dawkins himself have said fairer than all this?

The passages I have now quoted from *Adaptation and Natural Selection* are only a small fraction of those which could be quoted to the same effect. But they are probably enough to satisfy the reader that Williams is indeed engaged in explaining adaptation by the purposes of agents of super-human intelligence and power. Could *you*, or any other organism, *calculate precisely* how to enhance the reproductive prospects of the genes of an ancestor of the bird dropping spider, and then *actually* enhance them? No; but certain *genes* can, and they did. In short Williams, like Dawkins, differs from Paley only about the number of the gods responsible for adaptation, and about their moral quality: *not* about their existence, purposiveness, intelligence, or power.

Late in his book Williams, as though he felt he had still not done enough homage to the author of *Natural Theology*, goes out of his way to quote and praise a passage of Paley, on the subject of—of all shop-soiled examples!—the human eye. The passage is instructive, but too long to be quoted here.[23] I suspect that Williams wrote it partly for the purpose of shocking the duller witted, or more historically ignorant, of his fellow Darwinians.

Williams is lacking (as I have said) in the literary gifts which could have transformed the dry bones of his book into a "living garment" of the new religion. He left that task, perforce, to Dawkins. Williams never calls genes *selfish*. He never says that they *manipulate* the organisms that carry them, and still less that they manipulate everything else in sight, or out of it, for their own ends. He never talks about people or other organisms being "robots," or "survival machines" designed, built, and operated in the interests of "the selfish molecules called genes."

All the same, he did quite enough to "make straight in the desert a highway" for the new religion of selfish genes. He *does* say (as we have seen) that the adaptations of organisms are "tools and tactics for the strategy of genes," and that "the ultimate goal" of all adaptation is the continuance of the genes concerned.

He is equally "Dawkinsian" (to reverse the real order of things) on all the other subjects which agitate gene religionists: altruism towards non-kindred, for example. This, Williams peremptorily says, is either non-existent, or is a "biological error" where it does occur.[24] He even gives—though this fact is scarcely credible—as a *typical example* of what the relations among non-kindred conspecifics are in general like, the relations among "the house cat population of any neighborhood."[25] As though such animals are not universally recognized, even to a proverb, as exceptionally unsocial ones!

Truly, though Williams left much for Dawkins to do in the way of popularization, he left him singularly little to do in the way of intellectual substance. And in particular—the thing which principally concerns us in this essay—he sounded so loudly and insistently the Paleyan note, of the purpose, intelligence and power displayed in adaptation, that he could not have failed to be the inaugurator of a gene religion, wherever he was believed.

THUS HAS PALEY had his long-delayed revenge on Darwinism. For more than a hundred years, the proudest boast of Darwinians had been that they had at last complied with Bacon's famous injunction and expelled "final causes" from their science. Paley was remembered, when he was remembered at all, only as the most atrocious of all offenders *against* that injunction. And yet we find, in the last third of the twentieth century, many Darwinians of the highest reputation ascribing adaptation to the purposive activity of beings which possess more than human intelligence and power. This is certainly a sufficiently remarkable historical comeback; even if Paley *redivivus* has had to settle (as I said) for plural and immoral divinities.

Williams would of course deny that he attributes any purpose to genes. Dawkins likewise, and he has in fact expressly denied in print that he does explain adaptation by reference to purposeful agents. Both these writers claim to be, and certainly claim sincerely to be, firmly in the old Darwinian tradition: the tradition of explaining evolution in general, and adaptation in particular, by reference to *blind* causes, altogether devoid of purpose or intelligence.

Dawkins, in order to make clear the great *difference* between the Paleyan explanation of adaptation and his own Darwinian one, writes (for example) as follows. "Natural selection . . . has no purpose in mind. It has no mind and no mind's eye. It does not plan for the future. It has no vision, no foresight, no sight at all."[26]

These statements (though excessively repetitive) are all true. But alas, they are trivial. For they would still all be true, if we were to put for their subject, instead of "natural selection," "artificial selection." Artificial selection has no purpose in mind. (Cattle breeders have, though.) Artificial selection has no mind. It does not plan for the future (though wheat geneticists do). But no one would be tempted to infer, from these truisms, that purposeful intelligent agents play no part in bringing about *artificial* selection!

In fact the truth of the statements just quoted from Dawkins is a trivial consequence of his having chosen an abstract phrase, "natural selection," as their grammatical subject. In the same way, we could say, with equal truth, that (for example) "business competition" has no mind, or that "warfare" does not plan for the future. But it would be an exceptionally gross error, to infer from these trivial truths, that the purposes and intelligence of *businessmen* are not among the causes which determine the success or failure of firms, or that the purposes and intelligence of *soldiers* are not among the causes that decide which army loses and which wins.

The question is not, then, whether natural selection is purposeful. The answer to that question is trivially, no; just as it is for artificial selection, business competition, or warfare. Breeders, businessmen, and soldiers, however, certainly *are* purposeful and intelligent, and the right question to ask of Williams and Dawkins is, whether *genes* are purposeful? That genes are at least among the causal agents which bring about adaptation is agreed on all hands. The question is whether they are *purposeful* ones?

It ought to be unnecessary, but unfortunately it *is* necessary, to insist that this is not the same question as whether genes are *consciously* purposeful. Purposes do not need, either logically or empirically, to be conscious; and often they are not. Nor is this fact a "discovery" of Freud: it is a commonplace deliverance of common sense. People quite often realize that they have been, for some time, intending or "purposing" to bring a certain state of affairs about, without having been conscious at the time of having any such purpose. It cannot reasonably be doubted that much of the activity of dogs is purposive; but whether any of it is consciously so, may very reasonably be doubted. And purposes, of course, extend a long way down in the animal world below dogs, while *conscious* purpose can hardly even extend so far. "Purposeful" or "purposive" *does*, indeed, logically require "(at least minimally) intelligent," in the same homely but iron-hard way as "mermaid" logically requires "female," say, or "red" requires "colored." But "purposeful" does not logically require "consciously purposeful."

Dawkins has returned a clear "no," not only to the question whether natural selection is purposive, but to the question whether genes are so. Present-day genes, he says, "are no more conscious or purposeful than they ever were. The same old processes of automatic selection . . . still go on as blindly and as inevitably as they did in the far-off days. Genes have no foresight. They do not plan ahead."[27] And no doubt Williams too would say, if he were asked, that genes have no purpose.

What, then, is my excuse for saying, in this essay or in Essays 7 or 9, that Dawkins and Williams *do* ascribe purpose to genes? Why, a very simple and sufficient excuse. Namely, that for every once that Dawkins says that genes are not purposive, he says a hundred things (many of which I have quoted) which imply that genes *are* purposive. And that Williams, likewise, says countless things which imply that genes are purposive, although he doubtless believes (while never actually saying) that they are not. If the writer of a book says a certain thing twice or once or never, but implies the opposite over and over again throughout his book, a rational reader will take it that the writer's real opinion is the one which he constantly implies; not the other one.

That Dawkins and Williams *do* constantly imply that genes are purposive does not need to be proved now, because I have proved it already by many quotations from them which I have given in Essays 7 or 9 or in the present one. They actually refer in some of these passages (as we have seen) to genes as having "their own ends," or as having a "goal." But it would not matter even if they had never done so. For the same implication is clearly present in all the references which these authors make to the various "tools," or "tactics," or "devices," which genes employ. These references to the *means* that genes make use of imply purpose just as much as their references to the ends or goal of genes.

Manipulation by genes (of their carriers, of other organisms, of the environment, etc.) is the central conception of the new religion, as I pointed out in Essay 9. But the manipulation logically requires the presence of an intention or purpose. If there is no intentional causal influence, then there is no manipulation. And this is simply a transparent logical truth about the meaning of a common English word: just like (say) the truth that where there is no color there is no redness, or that no mermaid is present if no female is.

In Essay 7 I pointed out that Dawkins relies upon (though he sometimes disavows) the ordinary psychological sense of "selfish" and its cognate words. In that sense, a selfish person is simply an unusually or unduly self-interested one. But something which has no interests, intentions, or purposes at all, logically cannot be selfish. So when Dawkins calls genes "selfish," he implies that they have interests or purposes.

The distinction so much insisted upon by Williams, between the function of an organ (say) and effects which that organ has in the ordinary causal way, is one which we all understand quite easily. The distinction, for example, between the heart's circulating blood and its making a sound. At least, we all understand this distinction well enough, as long as we are allowed to think of a function in the way in which Williams himself always speaks of it. That is, as something *designed to* have a certain effect, or distinct from something which merely has that effect *de facto*. But how are we to understand the distinction when we are no longer allowed to think in that way, but are told instead that a blood-circulating heart (or any other adaptation) is *designed by genes*, and yet that genes have *no* purposes?

All of these are matters which I have sufficiently touched upon already. Is additional evidence wanted, that Dawkins or Williams imputes purpose to genes? Then I will mention the rivalry or competition or struggle between genes which are alleles of one another to increase or at least maintain their "market share" in the next generation of organisms.[28] Now, two things cannot be rivals of or competitors with one another, and cannot struggle against one another, unless at least one of them is *trying* to achieve something. We may say, indeed (for example), that the rocks of an exposed sea shore struggle against the waves and the winds. But this is only by courtesy of what everyone recognizes as a metaphor, since everyone knows that neither rocks nor waves nor winds have any purposes: they are not *trying* to do anything. By contrast, the struggle between two alleles is (as I have just said) a struggle *for* something, or *to* achieve something: namely, increased or not decreased representation in the next generation. Those little words "for" and "to" (when it is short for "in order to," as it is here) are of course the tell-tale and indelible signs of purpose of intention.

Dawkins told the readers of *The Selfish Gene* that, if they objected to his describing genes as selfish, he could easily "translate [that statement] back into respectable language."[29] Well, I do object to it, and one of the grounds on which I object to it is that it implies that genes are purposive. So I would like to know what the "respectable translation" is of "genes are selfish."

It would be—this much is clear—some statement about the propensity of genes to replicate, and it would *not* imply that genes are purposive. But this does not tell us what the translation actually is. Dawkins himself does not tell us what it is. So we will have to try to work it out for ourselves.

The natural first candidate to consider is just: "genes are replicators." But this will not do. If it means that all genes replicate, it is false straight off, since many do not; those of childless people, for example. If it is true, it can only be in the indefinite way that "dogs bark," "fish swim," and "cats are fish eaters," are true. But then, as a translation of "genes are selfish," it is altogether *too* respectable; in fact, an embarrassing anticlimax. "Genes are replicators" gets rid of the implication of purpose, all right. But ". . . are replicators" falls so very far short in meaning of ". . . are selfish" that the translation is a truth not worth stating. *Everyone knows* that genes are replicators. But not everyone knows (to put it mildly) all the hair-raising consequences which follow at once from the proposition that genes are selfish.

We might try: "Genes are replicators, and every gene does replicate if it gets a chance." This is certainly closer than "genes are replicators" to what Dawkins meant by "genes are selfish." The trouble is, it is too close to be respectable. For it clearly reintroduces the idea that genes are purposive: that they *try* to maximize the number of their replicas. But that is the very implication which we are trying to *translate out* of "genes are selfish."

In an attempt to avoid that implication, we might consider instead: "Under an extremely wide range of circumstances, genes *do* replicate." But as a translation of "genes are selfish" this is, like "genes are replicators," altogether too respectable and uninteresting. The world is full, after all, of things which happen under an extremely wide range of circumstances: heartbeats making a noise, bullfrogs croaking, birds getting parasites, and businessmen getting ulcers. But no one could ever suppose that these are the *biologically central* things about heartbeats, bullfrogs, birds, or businessmen: the things which explain everything else about those entities. Whereas "genes are selfish" *is* supposed to pick out the biologically central thing about genes, which explains everything else about them.

What about this: "Genes do replicate under an extremely wide range of circumstances, and they don't do anything else"? Well, this suggestion is definitely in the *spirit* of Dawkins' "genes are selfish." But of course it is hopelessly false. Molecular biologists can tell you a hundred other things that genes do besides replicating.

Here is another suggestion: "Genes do replicate under an extremely wide range of circumstances, and everything else that they do is for the sake of replicating." And now we really are getting "warm" (as children say). But alas, this translation is *too* close to "genes are selfish," and as a result it lacks the promised respectability. For it clearly reintroduces the very thing that we are trying to translate out: the implication that genes are purposive. Talk about certain things being done "for the sake of" something or other is plainly just as teleological as talk about ends, goals, or purposes.

It is certainly no easy matter, then, to cut out of the statement that genes are selfish, the implication that genes are purposive, while leaving that statement both true and worth making. I am not suggesting that such a "respectable translation" of that statement is *impossible*. I do not know whether it is possible or not. But I do say that neither Dawkins, nor anyone who substantially agrees with him, has actually given this translation which they promise us; that neither I nor (as far as I know) anyone else knows what the translation is; and that Dawkins and those who agree with him must be *able* to give such a translation, if they are to escape the accusation that they ascribe purpose to genes.

Of course it is not just the statement that genes are selfish of which Dawkins owes his readers a translation into respectable non-purposive language. He equally owes us similar translations of all his countless statements about "manipulation" by genes, about the "tools" and "tactics" they make use of for their own "ends," about the "rivalry" between alleles "for" a place on the chromosome, and so on. Williams, similarly, owes us a translation into non-purposive language of all his innumerable references to adaptations as things which don't just happen in the ordinary causal way, but are *designed*. Even if Williams has forgotten the fact, it *is* a fact about the meaning of a common

English word that you cannot say that something was designed, without implying that it was intended; any more than you can say that a person was divorced, without implying that he or she was previously married.

Indeed Williams, Dawkins, and those who agree with them, owe the rest of us a whole *translation manual*: a manual which will tell us how we are to understand all the statements they make which, if they are understood in the usual sense of the words in them, imply that genes are purposive. Until such a manual is available, selfish gene theorists cannot reasonably complain if other people regard them as just propagandists of yet another new religion. Human life swarms, after all, and always has swarmed, with groups of people who claim to have fascinating news to impart, concerning purposeful beings of more than human intelligence and power; and *a priori* it is perfectly possible (as I said in Essay 9) that the claims of one of these groups are simply and literally true. Rational people, however, treat all such claims with extreme caution. But when a certain group of people make claims of this sort, and at the same time give themselves out as accepting the *Darwinian* explanation of adaptation, then a rational person will exercise a double dose of caution. At the very least he will *ask to see their translation manual*, so that he can satisfy himself as to what they really mean, when they describe genes as designing, manipulating, competing, being selfish, etc. That they do not mean what they *say*, we know both from their own admission and from the Darwinian explanation of adaptation. But what they *do* mean, they do not tell us. No translation manual exists, or even the beginnings of one.

This situation is evidently unsatisfactory; but it is far from being new. In fact, in its essentials, it has existed ever since 1859. Darwinians have *always* owed their readers a translation manual that would "cash" the teleological language which Darwinians avail themselves of without restraint in explaining particular adaptations, into the non-teleological language which their own theory of adaptation requires. But they have never paid, or even tried to pay, this debt.

Darwin, for example, published in 1862 a book entitled *The Various Contrivances by which Orchids Are Fertilized by Insects.* He knew, and all his readers knew, that he did not really mean the word "contrivances." Everyone understood perfectly well (a) that you cannot call something a contrivance without implying that it was intended, and (b) that Darwin did *not* mean that these "contrivances" of orchids were ever intended by anything.

He therefore owed his readers an explanation of what he *did* mean by "contrivances": a translation of that word into language free from the implication of intendedness. But he never gave such an explanation or translation. Since, presumably, he would have done so if he could, I suppose the reason was that he did not know how to. Nor have any Darwinians ever given, to this day, any such reconciliation of their theory with the teleological language which they employ as freely as though they were disciples, not of Darwin, but of Paley. Presumably the reason that they have not is the same as the reason Darwin did not.

I am not suggesting that Darwin *should not* have used, or that a Darwinian should not use, teleological language when trying to explain particular adaptations. That would be a hopelessly doctrinaire and impracticable suggestion. A biologist, whether of Darwin's time or ours, can hardly frame a single thought, concerning adaptations, which does not involve intendedness or purposefulness. To ask him to purge his mind of all such thoughts, and never to use worlds like "purpose," "function," or "contrivance," would amount in practice to telling him to stop thinking about adaptation altogether.

I DO SAY, though, that Darwinians cannot reasonably expect, any more than anyone else can, to be allowed to have things both ways. They cannot, on the one hand, describe adaptations as contrivances for this or as designed for that while denying that they mean that these adaptations were ever intended; and on the other hand, decline to explain what they *do* mean by expressions like "designed for" and "contrivance for."

Darwinians, then, have never paid, or even acknowledged, the debt they have all along owed the public: a reconciliation of their teleological explanations of particular adaptations with their non-teleological explanation of adaptation in general. And not only have they never paid this debt: they have in fact become progressively less conscious, with time, of the fact that they owe this debt. This is a natural failing, of course, in people with debts which have remained unpaid for a long time. But it is not the less, on that account, an inexcusable failing. Intellectual debts (whatever may be the case with economic ones) are not extinguished merely by being ignored or forgotten, for however long a time. There is no Statute of Limitations which says that Darwinians may—as long as they go on doing it *long enough*—imply that adaptations are intended, and *say* that they are not.

In this respect, Williams is perfectly typical of present-day Darwinians. He must have known, at the time he first became a Darwinian, that he used an expression like "was designed for" with an invisible promissory note attached to it, saying something like "To be cashed at a later date in non-teleological terms." By the mature age at which he comes to write *Adaptation and Natural Selection*, however, he has, like any other mature Darwinian, issued so many of these promissory notes, that he is no longer conscious of their existence. He has simply forgotten what teleological words *mean*, or else has forgotten the fact that they are not really available to Darwinians engaged in explaining adaptations. In particular, he has forgotten that "was designed for" implies "was intended to." But unluckily for Williams (though luckily for sanity and for non-Darwinians), "was designed for" still means what it meant before the Revelation of 1859, and in particular, still implies "was intended to." And that being so, Williams does still owe his readers the translation of his talk about design into non-teleological language which he, in common with all other Darwinians, has been promising for so long, and yet never performed.

Although Darwinian biologists have never tried to discharge the intellectual debt they owe in connection with adaptation, there is a certain group of *friends* of Darwinism who have behaved more conscientiously. I refer to the many philosophers who have discussed teleology in the last fifty years, and attempted to provide for Darwinians the translation manual which they have always needed, but never tried to provide for themselves. These writers *have* considered, and in a most searching way, whether explanations of adaptation *can* be purged of every implication of purpose, and thus reconciled with Darwinism.

This literature is by now extensive, and has become somewhat specialized. But any educated person can form a fair impression of it, by reading for example Chapter 12 of Ernest Nagel's *The Structure of Science* (1961),[30] Andrew Woodfield's *Teleology* (1976),[31] and Chapter 1 of Alan Olding's *Modern Biology and Natural Theology* (1991).[32] Any Darwinian biologist who suspects that what these authors are trying to do does not need to be done, or that he could do it as well or better himself, would benefit by reading these writings.

Nor have these enquiries of philosophers been without results: quite the reverse. But it must be admitted that all their results have been negative. One philosopher proposes a non-teleological "analysis" of (say) "The function of the heart is to circulate blood," which looks watertight, but then a critic points out that the proposed analysis is insufficient or unnecessary for the truth of "The function of the heart is to circulate blood," or that it contains a covert reference to purpose, or to another concept of the same family. This has by now happened very many times. It has turned out, in fact, to be far harder to translate teleological into non-teleological language than had been anticipated by philosophers; or at any rate, by philosophers friendly towards Darwinism (as virtually all the writers in question are). Whether such translation is possible at all is more than anyone knows.

As I HAVE SAID, no Darwinians have ever been interested in providing the translation manual that they should provide; but easily the *least* interested of all Darwinians in providing that manual are the adherents of the new religion of genes. These people have *found* their divinities, and know how to describe them. They have long since passed the stage at which any merely *human* criticism can trouble them. Indeed, even if the Supreme Gene of all genes were itself to say to Richard Dawkins, "You have, perhaps, somewhat overestimated the intelligence and power of us genes," it would not do any good. Dawkins would at once recognize this communication as just another instance of selfish manipulation by genes, and would merely feel surer than ever that he had *not* overestimated the intelligence and power of those superior beings.

It is very easy to understand why gene religionists are especially uninterested in the project of a translation manual. For suppose that that project were successfully completed: what would be the effect on the new religion? Simply that Dawkins and like-minded people could no loner describe genes as selfish, as manipulating everything under the sun, as competing with their alleles for market share in future generations, and so on, *while promising—but never performing—a translation of those statements into "respectable language."* There would be no need any longer, and no excuse, for describing genes in ways which imply, as those do, purpose, intelligence and power. Everything that Darwinians needed to say, in order to explain adaptation in particular, could then easily be said, and would be said, without expressions such as "selfish," "manipulation," or "competition for." But this is to say, of course, that the new religion of genes would simply vanish like a dream, and "leave not a wrack behind."

IV

Organisms have the adaptations that they do, according to the religion of Paley, because a single benevolent God intends them to survive and reproduce; and because that intention will be fulfilled the better, the better adapted the organisms are. According to the new religion, organisms have the adaptations they do, because many selfish gods intend to have copies of themselves, and as many copies as possible, carried by the next generation of organisms; and because that intention will be fulfilled the better, the better adapted the organisms are.

Now, why is it that the idea of intention keeps turning up in explanations of adaptation, intruding even into ones where it is supposed to have no place? And why is it as hard, as we saw in the preceding section that it is, to translate the idea of intention *out of* the explanation of any particular adaptation?

"Surely it is just because any adaptation strikes us—as you yourself said in connection with bird dropping spiders—as a clever idea which has been intentionally and well carried out?" Not quite, for it is not true that adaptations strike *everyone* in that way: it depends on what *else* they believe. Mimicry in spiders of a bird dropping will not strike you as a brilliant idea, unless you think of someone or something as *intending* that those spiders should elude predators, capture prey, and in general, should survive and reproduce. It is only as a *means to that end*, that this mimicry looks like a clever idea well put into practice. You may think of the intending as being done by the spiders, or by God, or by the spiders' genes, or by whatever. But if you do not think of *anything at all* as intending these spiders to survive and reproduce, then their resemblance to a bird dropping will not strike you as a brilliant idea. It will not strike you as anything in particular, except as an odd coincidence.

THE MOST NATURAL thing to think, of course, is that it is the spiders themselves who intend that they should survive and reproduce; and that is what, nowadays, everyone does think. Or rather, we take it absolutely for granted, and we have taken it so for several centuries. That is why the resemblance of these spiders to a bird dropping strikes us as a great feat of intelligence and engineering. But I venture to affirm that before the modern period—before 1600, say—no one, or virtually no one, ever thought of organisms in that way. People must always have known, of course, that beasts of the chase, and the weeds and insects which harm crops, do not surrender their lives at our request. But the general conception, of all organisms as striving to the utmost to survive and reproduce, seems never to have existed in antiquity or the middle ages.

So I say (for example) that if Aristotle had discovered the bird dropping mimicry of spiders for himself, he would *not* have thought, as any modern person would, "What a clever idea for capturing prey and eluding predators!" *We* think that, and have thought so for centuries, partly because we have for centuries taken it for granted that spiders are intent on, indeed fanatically intent on, surviving and reproducing. But I do not believe that Aristotle, or anyone else before about 1600 A.D., ever thought of organisms as intending our purposing to survive and reproduce, let alone as being inflexibly bent on that goal.

Thinkers of antiquity or the middle ages did, of course, often postulate purposes in order to explain certain natural phenomena. Indeed, they postulated purposive causes far too freely (as is well known). With Aristotle, even physics is teleological. Astronomers, when they were Christians, were always sure to postulate divine purposes in order to explain some feature or other of the cosmic layout. But purpose was *not* postulated to explain the one thing which, to all modern minds, seems most manifestly to require a purposive explanation: the survival and reproduction of the countless species of organisms that we find around us.

Does this assertion appear incredible? Then I will point out an astounding fact which will go far towards making it credible. This is, that although design arguments for the existence and purposes of God are at least 2,400 years old, virtually no one before the seventeenth century ever based a design argument on *the adaptation of organisms*. In fact (as far as I know), only one person ever did: Galen, the great doctor and medical writer of the second century A.D., who laid the foundations of human anatomy.

In the seventeenth and eighteenth centuries, indeed, the design argument based on adaptation "ran riot," and pushed every other special form of that argument into the background. By 1800, adaptation had become not merely the main, but virtually the only empirical evidence appealed to, to establish the divine existence and purposes. Paley sufficiently indicates that he himself attached little value to the design argument when it is based on anything *other than* adaptation.[33] And yet when Plato or Aristotle or Cicero or Aquinas had employed a design argument, it had never been from adaptation. It was always from some fact, or supposed fact, of astronomy, or of general or terrestrial physics: from almost anything in the world, in fact, *except* the adaptations of organisms.

Now, where there is no recognition of the universal striving of organisms to survive and reproduce, there can be *a fortiori*, no recognition of a further fact: that in organisms in general, all other purposes—to establish a territory, to utter a certain call, to intimidate a rival, etc.—are *subordinated* to the overarching purpose of surviving and reproducing.

To modern minds, again, this fact is a complete commonplace, and has been so for centuries. The fact is subject to an important exception (I need hardly say) in the case of man: human beings have many purposes which are *not* tributary to the purpose of surviving and reproducing, and many which even conflict with that purpose. But outside man and the few animals and plants he has domesticated, the generalization holds profoundly true. Everything that other organisms do, or try to do, *is* subordinated to the goal of surviving and reproducing. But you will search in vain for knowledge of this fact in antiquity or the middle ages.

Further: where there is no recognition that all the other purposes of organisms subserve their arch purpose of survival and reproduction, there can be no recognition, *a fortiori*, that for organisms in general, it is *difficult* and *dangerous* to survive and reproduce. There can be no recognition that life in general is a *struggle* to survive and reproduce; and a struggle at that, without pity, without exemptions, and without end. To modern minds, again, this proposition is a commonplace, and has long been so. But where is the conception of life, as always and everywhere a struggle, to be met with in antiquity or the middle ages? Again I venture to say, nowhere. The idea that *human* life had once been a "war of each against all" is ancient, indeed. But the idea that *all* life *is* a struggle to survive and reproduce is not ancient, or even old.

Further still: where there is no recognition that organisms have to struggle in order to survive and reproduce, there can be no recognition, *a fortiori*, that because of the exuberant tendency of all species to increase in numbers, a large part at least of their struggle for life must always be a struggle *with their conspecifics*.

With us, yet again, this recognition has long been something taken for granted. But (as I said in Essay 2) recognition of the pressure of population on food supply, as a *general* biological fact, appears to be only about fifty years older than Malthus's *Essay* of 1798. Before the generation of Benjamin Franklin and David Hume, no one seems ever to have grasped it at all. Of course Plato, Aristotle, and other actual or would-be rulers of states had realized the danger to stable government which un-restrained increase of human numbers can present (as Malthus reminded the readers of his *Essay*). But that is an entirely dif-ferent thing from realizing that, in all species of organisms, population always presses on the "means of subsistence."

These truths—that organisms strive to survive and reproduce, that all their other purposes are subordinated to that one, that they are obliged to struggle for life and to struggle largely with their conspecifics—are all elements, of course, of the Darwinian explanation of evolution. And since they are so, there could hardly be a greater mistake than a certain statement which has been endlessly repeated: namely, that whereas the Lamarckian explanation of evolution accords a causal role to the purposes of organisms, the Darwinian one does not.

This is, in fact, a misrepresentation of the Darwinian theory so great as to make any rational person marvel how it can ever have acquired currency. The famous Darwinian "struggle for life," on which the whole theory turns, is a struggle *for* something, is it not?: namely, for survival and for leaving descendants. But in that case it is a *purposive* activity on the part of the individuals which struggle. And in any case, Darwin is always saying things like the following: that "each organic being is *striving to* increase at a geometrical ratio";[34] or that "every single organic being around us may be said to be *striving to the utmost to* increase in numbers."[35] How could he have ascribed purpose to all organisms more plainly than this? Nor has anyone, even the most behavioristic or positivistic Darwinian, ever credibly suggested that *these* references to purpose can be, or need to be, expunged or "translated out" of the Darwinian explanation of evolution.

In fact it is precisely the striving of organisms to live, reproduce, and increase which, according to Darwin, *drives* the whole gigantic process of evolution. If organisms were indifferent towards their own survival and reproduction, or if they positively leaned to the Buddhist side of those issues, there would be no struggle for life, hence no natural selection, and hence no evolution, according to the Darwinian theory. So very far is that theory, then, from according no causal role in evolution to purpose.

For this same reason, we should not let ourselves be imposed upon by another group of commonplaces: the ones about Darwinism having expelled "final causes" from biology. If "final causes" means purposes, or purposive activities, then Darwinism not only does not "expel" them: it builds them into the very foundation of its explanation of evolution.

Even the common statement (which earlier in this essay I repeated myself) that Darwinism explains evolution solely by reference to "blind" causes can be accepted only with certain reservation. The statement is perfectly true, if "blind causes" just means "non-conscious causes." But (as I pointed out in Section III), "not conscious" does not imply "not purposive": for purposes need not be conscious. And if "blind causes" means "causes which are not only not conscious but not even purposive"—causes like gravitation, or friction, say—then it is simply false that Darwin explained evolution solely by reference to blind causes. For one of the causes by reference to which he explained evolution was the striving or intention or purpose of organisms to survive and reproduce.

It must be admitted that Darwin's language always starts to

display a marked hesitancy or embarrassment the moment he ascribes a purpose to all organisms. The two quotations given a few paragraphs back furnish an example of this hesitancy. Why is it "striving" in one of these passages, and "*may be said to be* striving" in the other? This feature of Darwin's language reaches a climax in the fourth paragraph of Chapter III of the *Origin* (entitled "Struggle for Existence"). That paragraph is a long and uneasy discussion of *exactly* when organisms may "properly . . . be said," or "may truly be said," to struggle for life. But in fact almost every time that Darwin ascribes striving or struggle to organisms in general, you will find that the qualification "*may be said to be* [striving or struggling]"[36] is prefixed.

But all this hesitancy is no reason whatever to doubt that Darwin does ascribe to all organisms at least one purpose: that of surviving and reproducing. His embarrassment is sufficiently accounted for by two difficulties which beset *everyone* who writes about purpose in nature: difficulties which no one, to this hour, has been able to solve. One difficulty is that on the one hand we are reluctant to ascribe purpose to plants and to lower animals, while on the other, their behavior compels us to describe them as striving or struggling or trying to survive and reproduce.

The other difficulty is that on the one hand it seems to be stretching matters somewhat to say that organisms strive to *increase*; and even (when you stop to think about it) to say that they strive to reproduce. What they do, beyond all question, strive to do (apart from surviving) is to *mate*. But then, on the other hand, the causal connection between their mating and their reproduction and increase is so extremely intimate and inevitable that—well, after all, the best solution *is* to say that organisms do strive, not merely to mate, but to reproduce, and even to increase. It is, admittedly, "telescoping" different things to say so: it is pushing together an intended activity, mating, and its unintended consequences. And yet, to *separate* mating from reproduction, or reproduction from increase, where organisms in general are concerned, only involves us in unmanageable complexities of thought.

Both these difficulties, as will be evident, are imposed upon us by real continuities—either qualitative or causal continuities—in organic nature; and if Darwin did not satisfactorily solve the difficulties, that is no more than can be said with equal truth of everyone else who has ever wrestled with the subject of purpose in nature. The two difficulties *would* be solved, indeed, if we were able to accept the "pan-psychism" which has been advocated by Leibniz, Samuel Butler, and A. N. Whitehead among others. That is, if we could believe that purpose belongs to every last element, not only of organic but of inorganic nature. But to most people, this seems to be a case in which the cure proposed is worse than the disease.

That organisms strive to survive, reproduce, and increase; that any other purposes they may have are subordinated to those great ends; that organisms have to struggle in order to achieve even the first of these objects, survival; and that a large part of their struggle for life is with members of their own species: these four propositions, then, though they have long been common knowledge in the modern period, were unknown at any earlier time.

Most people nowadays would associate these four propositions with Darwin, and imagine that we owe our knowledge of them to him. But this is merely from ignorance: the historical fact is exactly the opposite. Darwin found all of this knowledge ready to his hand when he first began to study biology. The four propositions just mentioned were quite as well known to Paley in 1802, for example, as they were to Darwin in 1859.

Paley knew perfectly well the central place occupied by the sexual reproductive impulse in life in general. He was vividly aware of its overwhelming strength, not only in animals, but in plants. See the whole of his excellent Chapter XX, "On Plants," especially on the care taken by the parent plant to develop, protect, ripen, and finally disperse, that "sacred particle,"[37] the seed. Then, Paley had read and absorbed Malthus's *Essay*, published four years earlier, and was fully conscious of the "superfecundity"[38] of all organisms. And he saw clearly that the consequence of their superfecundity must be that the members of every species will always "breed up to a certain point of *distress*,"[39] and be obliged to struggle for life against their conspecifics; with most of them achieving but little success in that struggle.

Knowledge of the four propositions I have mentioned ought to lead anyone to recognize that life in general is principally a scene of care, effort, anxiety, pain, disappointment, and early death. But Paley was also, after all, one who believed in the benevolence of God. Hence his painfully unconvincing attempt to prove, in the teeth of his own better knowledge, that "It is a happy world after all."[40] Even his most plausible examples in support of this proposition were of a kind ready at a touch to undermine his avowed optimism; for these examples were drawn (as he himself remarks) from the *young* of various species. A profounder thinker than Paley, though one equally without benefit of Darwinism—Schopenhauer—was later to point out the main reason for the comparative happiness of the young: the fact that the weight of their mirthless biological destiny has not yet descended upon them.[41]

The pressure of population on food, and the struggle for life among conspecifics which results from it, *were* only recent discoveries (as I have said) when Paley wrote *Natural Theology*. But the same is not true of the first two of the four propositions mentioned above: that organisms strive to survive and reproduce, and that their other purposes all subserve that one. When *these* biological truths were discovered, or by whom, it appears to be impossible to learn, except that it was during the seventeenth and eighteenth centuries. They seem never to have been "discovered," in any distinct sense, at all, but rather to have diffused themselves imperceptibly over the minds of all naturalists, as their studies of plants and animals became both more extensive and accurate, and, late in the eighteenth century, more systematic as well.

But there *was* one identifiable discovery which did contribute powerfully to the recognition that all organisms strive to reproduce: the discovery of the sexuality of plants. That flowering plants reproduce by the intercourse of organs of different sex had been known to that astounding genius Empedocles in the mid-fifth century B.C.; and Theophrastus, Aristotle's successor late in the fourth century, shows acquaintance with some agricultural practices which imply knowledge of the fact that plants reproduce sexually.[42] But after that, the sexuality of plants faded out of human knowledge for fully 2,000 years. It began to "fade in" again, late in the seventeenth century and early in the eighteenth, first through the researches of Nehemiah Grew (1641–1712), but more importantly through those of Rudolf Camerarius (1665–1721). Even then, knowledge of the sexuality of plants spread only slowly, and especially slowly outside the ranks of professed naturalists. Paley had fully absorbed it, and even regarded "the sexual system" of reproduction as sufficient on its own to establish the benevolent purpose, intelligence, and power of God.[43] But then Paley was an exceptionally well-informed man. Some other educated Englishmen, as late as 1798, still thought that the sexuality of plants was a lie, and just another Jacobin attack on morality and public order. In that year, Erasmus Darwin's exposition of the sexuality of plants, in his long and mildly salacious poem *The Loves of the Plants*, was parodied and ridiculed in the *Anti-Jacobin* magazine.[44]

Whether the man in the street, even now, can properly be said to know of the sexuality of plants may very well be doubted. And yet few discoveries were of more importance towards making possible a true conception of life. For one thing, it revealed a previously unsuspected identity, and an identity of the deepest kind, between life's two great divisions: animals and plants. Second, and even more importantly, it dealt a tremendous blow to our anthropocentrism, which had previously defied all attempts at cure, and which the biology of antiquity or the middle ages had never even attempted to cure. For it revealed that wheat, apple trees, roses, and oaks do not (as we find it so natural to think) exist for *our* sustenance, delight, or use: that on the contrary, they have a purpose of their own, an overriding purpose too, and one which they share with all other organisms—to survive and reproduce *themselves*.

But the discovery of the sexuality of plants was not only *intellectual* dynamite: it was moral and political dynamite as well. For the Christian religion, after all, had waged war from its very start against the sexual impulse in man: not just against its hypertrophy, but against the thing itself. It had always been obvious to every thoughtful person that the sacrament of Christian marriage was no more than an uneasy compromise with the deadly sin of concupiscence. And yet, how could something which not just we and the "beasts" do, but which wheat and apples and roses and oaks do, be an offence against the divine nature and purpose? The conclusion which was bound to be drawn, and was drawn, was that, in spite of St. Paul, *sexual intercourse is innocent.*

This was a conclusion, of course, which Enlightened persons of the eighteenth century were already becoming convinced of on other grounds: discovery of the sexuality of plants merely provided the final scientific proof of it. Universal sexual emancipation had been high on the agenda of the Enlightenment from the start, along with the destruction of religion and of monarchical government. Sexual intercourse was to be freed, in the happy future, from the trammels of religion and "priestcraft," and of all laws of marriage, property, or inheritance. Contraception and feticide, from having been offences against divine, or moral, or civil law, were to be made not merely universal rights, but civic duties.[45] The great sexual emancipators after 1859—Havelock Ellis, Freud, Lenin, Marie Stopes, Margaret Sanger, Margaret Mead, Wilhelm Reich—were all Darwinians as a matter of course. But heroic labors in this great cause had been performed earlier, by the evolutionists Diderot and Erasmus Darwin, by Condorcet, Godwin, Shelley, and Fourier among others. Fourier looked forward to the day when Europe would be permanently crisscrossed, no longer by armies or crusaders or missionaries, but only by troupes of the most renowned sexual athletes, in permanent public competition, in order to keep before the eyes of the populace a constant reminder of the highest point which human felicity can reach.[46] When we, the beneficiaries of all these liberators, remember them in our grateful thoughts, we ought not to forget what they all owed, and therefore how much happiness we all owe, to the scientific labors of Camerarius and Grew.

IN FACT, THEN, the four propositions mentioned above, which we nowadays think of as together constituting the Darwinian conception of life, were all firmly established before Darwin was born. The following two quotations will suffice to prove this. For they are from Hume's *Dialogues* of 1779, and they clearly combine all the four components in question: the universal purpose of survival and reproduction, the overridingness of this purpose, the tendency of all species to exuberant increase, and the resulting struggle for life among conspecifics. (In both paragraphs the speaker is Philo, the character in the *Dialogues* who most closely represents the opinions of Hume himself; Cleanthes, whom he addresses, has been defending the design argument from adaptation.)

> You ascribe, Cleanthes, (and I believe justly) a purpose and intention to Nature. But what, I beseech you, is the object of that curious artifice and machinery, which she has displayed in all animals? The preservation alone of individuals and propagation of the species. It seems enough for her purpose, if such a rank be barely upheld in the universe, without any care or concern for the happiness of the members that compose it. No resource for this purpose: no machinery, in order merely to give pleasure or ease: no fund of pure joy and contentment: no indulgence without some want or necessity accompanying it.[47]

> Look around this universe. What an immense profusion of beings, animated and organized, sensible and active! You admire this prodigious variety and fecundity. But inspect a little more narrowly these living existences, the only beings worth regarding. How hostile and destructive to each other! How insufficient all of them for their own happiness! How contemptible or odious to the spectator! The whole presents nothing but the idea of a blind Nature, impregnated by a great vivifying principle, and pouring forth from her lap, without discernment or parental care, her maimed and abortive children![48]

THERE IS NOTHING in these passages (as will be evident) which bears at all on *evolution*, or on its explanation. But I am not here concerned with that. I am concerned with a certain *general conception of life*, which could be held whatever one believed the explanation of evolution is, or even if one had never so much as heard of evolution. And I cite the above passages in order to prove, by example, that the conception of life, which we (rightly) think of as Darwinism, owes nothing to Darwin, since it antedates him.

At some other places in Hume's *Dialogues* there *are*, as it happens, proto-Darwinian hints, both of evolution as an historical fact, and of the explanation of it by natural selection. But Hume, it is clear, was little interested in questions about what a given species, or a given characteristic of a species, has evolved from, or about how it evolved from it. And it must be admitted—even if the admission scandalizes Darwinian ears—that all questions of that kind are of little or no intrinsic interest.

The reason is that they are altogether too like certain other historical questions which are, by general consent, among the most boring ever propounded. Questions, I mean, such as "Where did the Hittites/the Maya/the Celts (etc.) come from, and how?" These questions are peculiarly pointless, because if the true answer to one of them *were* found, there would be as much reason as there was at first, to ask another question of exactly the same kind. If we learnt (for example) that the Celts came from place P, and did so by means M, how much would have been gained by this knowledge? We would have just as much reason as we had before, to ask where the Celts came *to place P* from, and by what means.

The same kind of uninterestingness attaches to all questions of evolutionary history: to all questions about what this species, or that characteristic, evolved from, or about how it evolved from it. Our species (for example) and any characteristic of ours, evolved, if it did evolve, from something *else*, and did so by some means or other. But just how it did, or from exactly what, are questions of no general interest.

But it is very different with the question whether a certain general conception of life is a true one, or not true. Is life in general the kind of thing which Darwin, or Hume in the passages just quoted, says it is? Or is it the kind of thing that Descartes says it is? Or the kind that Aristotle says it is? *These* questions cannot fail to be of at least some interest to everyone who is capable of understanding them.

Another example of the Darwinian conception of life being held by someone who owed nothing to Darwin, and an example even more arresting than that of Hume, is furnished by Schopenhauer, in *The World as Will and Representation*.[49] He died in 1860—the year after *The Origin of Species* appeared—and it is unlikely that he ever so much as heard of Darwin. He appears to have regarded man as an evolved species,[50] but he evidently feels, like a person of sense, little interest in that question. If your wish is to know and say what kind of thing human life is, then it will not make the least difference to you *what* human beings evolved from, or how, or when, or whether they never evolved at all, but have existed for all eternity.

Schopenhauer's central theme is the universality and, among organisms in general, the overridingness, of the sexual reproductive impulse. He calls this impulse "the will to live," or "the Life Force." This impulse or will or force is *purposive*—nothing more so, or more effectually so—but *not conscious*.[51] Though he never refers to Malthus, and perhaps had not read him, no one could be more vividly convinced than Schopenhauer is, of the grinding and constant pressure, in every species, of population on the supply of food; of the hair-trigger readiness of population to increase, in particular, if the food supply gives it the smallest chance to do so. The struggle for life among conspecifics, which results from this unsleeping and untiring attempt to increase, is universal, constant and pitiless.[52] Not even the Hardest Men among Darwinians have ever portrayed the struggle for life more uncompromisingly than Schopenhauer does.

He had a great (and just) admiration of Hume's *Dialogues*, and was tolerably well read in standard biological works: Kirby and Spence's *Entomology*, for example, and Richard Owen's palaeontological researches.[53] But Schopenhauer's general conception of life appears to owe less to his reading than to his gazing steadily at the phenomena of life themselves, without the smallest help from religious optimism, or from any other intellectual anodyne, such as the belief that biological evolution is progressive.

Schopenhauer had a special aversion to English clergymen[54] (a class which included both Paley and Malthus) and also to natural theology,[55] which was principally a creation of members of that class. His own religious leanings were entirely to the side of Buddhism (if Buddhism can be rightly described as a religion). At any rate, every deepest inclination of his nature led him to regard the spectacle of life as a whole with disgust and horror. "How frightful is this nature to which we belong!"[56] And what Darwinian, if he speaks from the heart, and drops for a moment his aspiration to scientific detachment, will not say the same?

Though a good writer, Schopenhauer is a very diffuse one, and as a result, quotations from *The World as Will and Representation*, unless they are given at enormous length, can convey only a very inadequate impression of his conception of life. As well as that, his key phrases "the Life Force" and "the will to live" are tainted, for most present-day readers, by having reached us *via* Ibsen and G. B. Shaw: disciples (by way of Nietzsche) of Schopenhauer, but not to be for one moment compared with him in either intellectual depth or breadth. Those phrases come to most of us through a distorting medium of what I have elsewhere called *horror Victorianorum*,[57] and we are therefore tempted to make fun of them—quite unreasonably. Finally, there is the fact that Schopenhauer himself "took the edge off" his conception of life, by embedding it in a Kantian *idealism*, according to which the physical universe itself exists only "for" a conscious mind.[58] This was, of course, a gigantic sop thrown to anthropocentrism: to the very thing, that is, which Schopenhauer just despised when English clergymen were guilty of it. His idealism is also (as I have shown elsewhere)[59] a very silly and trivial business. Fortunately, however, it can easily be "peeled off" his conception of the organic world, and thrown away, as it deserves to be.

Yet despite these impediments, some of them accidental and some self-imposed, Schopenhauer is, after all, the true philosopher of Darwinism. He was so before Darwinism existed, and he is so still. It is therefore entirely fitting that G. C. Williams should have disclosed, *via* a reference he makes to Buddhism, a deep affinity (though perhaps not a conscious one) with the Philosopher of Pessimism. "Perhaps biology would have been able to mature more rapidly in a culture not dominated by Judeo-Christian theology and the Romantic tradition. It might have been well served by the First Holy Truth from the Sermon at Benares: "Birth is painful, old age is painful, sickness is painful, death is painful . . ."[60] I do not know whether Williams, when he wrote these words, was conscious of a thunderous voice from the grave, saying "I told you so, long before Charles Darwin was ever heard of!" But there certainly was such a voice.

THE CONCEPTION OF LIFE, then, which we rightly call Darwinian though it owes nothing to Darwin, is this. All organisms strive to the utmost to survive, reproduce, and increase; everything they do, and all their adaptations, are contributory to that end; and it is only (or near enough only) the limitedness of their food, and the struggle for life in which it embroils conspecifics, which prevents them increasing without limit.

Here is this conception of life in Darwin's own words.

In looking at Nature, it is most necessary to keep the foregoing considerations always in mind—never to forget that every single organic being around us may be said to be striving to the utmost to increase in numbers; that each lives by a struggle at some period of its life; that heavy destruction inevitably falls either on the young or old, during each generation or at recurrent intervals. Lighten any check, mitigate the destruction ever so little, and the number of the species will almost instantaneously increase to any amount. The face of Nature may be compared to a yielding surface, with ten thousand sharp wedges packed close together and driven inwards by incessant blows, sometimes one wedge being struck, and then another with greater force.[61]

This arresting image of "ten thousand wedges," and in fact the whole last sentence of the above paragraph, occurs only in the first edition of *The Origin of Species*: all the five later editions simply omit the sentence. Yet Darwin must have valued this image, because by 1859 he had had it in his mind for twenty-one years;[62] and it is, indeed a powerful one. Still, it seems deficient in one respect: surely the "surface of nature" ought to be thought of as constantly striving to *increase* in area? Without this, the hammer blows of the struggle for life do not seem to have an active and purposive *opponent* to contend against.

Life, according to this conception of it, though it is hardly anywhere conscious, is *purposive* through and through. Organisms always and everywhere strive tirelessly to increase, and strive for no end which is not subordinate to that one. Their numbers grind ceaselessly on the supply of food available to them, and nothing can ever terminate, interrupt, or even alleviate this pressure of population on food while the species in question survives at all.

In particular, no effect of that kind can ever be brought about by intelligence, or by consciousness. Indeed, according to this conception of life, there could be no greater error than to think of intelligence and consciousness as *external to* the struggle for life, or as a possible source of interference with it. On the contrary, intelligence and even consciousness are just some of the means which have evolved in certain species *for use in* the struggle for life, and for nothing else; just as, in certain other species, a hard shell, or fleetness of foot, or a certain kind of dentition has evolved. The intelligence of higher animals, and the consciousness of humans, are merely *other weapons* employed in the struggle for life, and are entirely subordinate to their possessors' striving to survive, reproduce, and increase.

The Darwinian conception of life is an application of an ancient philosophical idea: "the principle of plenitude," as A. O. Lovejoy called in it the classic book[63] in which he described many earlier applications of this idea. This principle says that the world is full—a plenum—in the sense that there are *no unrealized possibilities*. Whatever is possible is actual, and the way the world is is also, down to its very last detail, the only way it could have been.

It is even obvious—once one pauses to think about it—that the Malthus-Darwin principle of population, which is the central element of the Darwinian conception of life, is an application of the plenitude principle. For it says (as we saw in Essay 2) that wherever there is food for a possible pine, person, or cod, there already is an actual pine, person, or cod, or else there will be, as soon as the reproduction time of the species in question allows it. This does not at all mean, of course, that population cannot *increase*. There will in fact be more cod, or more people, in six or nine months' time, than there are at present, if, though only if, an increased food supply, or increased mortality, makes that increase of population possible. But there are, at every moment, exactly as many organisms of any species as there could be: there are *no unrealized possibilities of reproduction*. Darwin said the same thing in the paragraph quoted above: "Lighten any check, mitigate the destruction ever so little, and the number of the species will almost instantaneously increase to any amount."

Now the principle of plenitude, and every application of it, seems hopelessly implausible when one first hears of it. For there appear to be, both in the inorganic and in the organic realm, *countless* unrealized possibilities: ways that things could have been, though they are not or were not. Before there was any life at all on earth, the wind at a certain place and time could have been a little stronger or more southerly than it actually was, could it not? Julius Caesar, at the time of his assassination, could have had one more hair on his head than he did in fact, or one fewer. I did not buy bread at the shop this morning, but I could have. The mosquitoes last summer (as someone actually said at the time) "could have been worse." You and your wife, who have had (let us suppose) three children, could have had just two, or one, or none. Unrealized possibilities seem, then, to be as common as dirt.

But against that there has to be set this fact: that the number of unrealized possibilities a person believes in seems to be strongly correlated with his or her *ignorance*. A child certainly believes in many unrealized possibilities which an adult does not believe in: princes being turned into toads, and so forth. An uneducated adult believes in many unrealized possibilities which an educated one does not believe in: miraculous cures of disease, or certain numbers being "lucky," etc., etc. In countless cases we find it natural to believe that a certain thing is possible, even if it never happens in fact. But growing up, or receiving some education, or scientific progress, constantly teaches us better: convinces us that what we had taken for an unrealized possibility is not possible at all, and is unrealized for the simple reason that it *is* impossible. The sphere of what we consider unrealized possibilities contracts, and with it, the number of things we consider *humans* free to do and capable of doing. We learn the chastening lesson that the world is, if not a plenum, at least closer to being a plenum than we had previously realized.

Many examples of this process, examples which were destined to become highly influential, were furnished by seventeenth-century physics. We are not usually conscious of the force of gravity, or of inertia, or of atmospheric pressure. As a result, we imagine that two bodies, left to themselves, could remain the same distance apart for all eternity; that a rolling billiard ball could change direction "of itself"; and that we could build a suction pump which would raise water more than thirty-three feet. We *can* build one which raises water more than twenty-three feet, after all: why not more than thirty-three?

But then along come Torricelli, Pascal, Galileo, Newton, etc., who make us better informed. They teach us that none of those three things is an unrealized possibility but that each of them is, on the contrary, unrealized because it is not possible. They thus bring home to us that the world is, at any rate, closer to being a plenum than we had thought before. The sphere of what we consider unrealized possibilities contracts, and with it, the sphere of what we consider ourselves free to do and capable of doing. We learn (for example) that we *cannot* build a suction pump which will raise water more than thirty-three feet. "The mechanism of the world picture"[64] takes a long step forward.

If this process of the mechanization of the world picture were carried to completion, it would mean, of course, that *all* our beliefs, about what we are free to do and capable of doing, are false. There would be *no* unrealized possibilities in life, any more than there are unrealized possibilities in arithmetic. It would turn out, since I did not buy bread this morning, that I could not have done so; that the mosquitoes last summer were as bad as they could be; and that you and your wife could not have had fewer children than the three you actually had.

Now, it is precisely this complete mechanization of the picture of life which Darwinism has always hoped to accomplish. This hope seems groundless at first, because we are not usually conscious of any universal striving of organisms to increase, or of any struggle for life which results from that striving plus the limitedness of food. For this reason, we are apt to think that *we*, at least, are not subject to those forces, but are free to act in ways other than the ways which they would oblige us to. But then, as Darwinians rightly remind us, we are not usually conscious of gravitation, inertia, or the pressure of the air, either, and had therefore been apt to think, before the seventeenth century, that various things are possible, and even within human power, which the physical discoveries of that century had proved to be nothing of the kind. The freedoms and powers with which we had credited ourselves, and the unrealized possibilities we had believed in, were merely the offspring of our ignorance and vanity.

In the same way, Darwinism says, biological science will in the end dispel *all* illusions of our being free and able to act otherwise than we do. We do not *feel* the universal striving to increase, or the struggle for life, any more than we feel gravitation, inertia, or air pressure, and yet the former forces really do constrain us just as rigidly as the latter do. The striving to increase, in our species as in every other, never sleeps, never tires, and never neglects an opportunity for reproduction. It is as constant, as irresistible, and as impervious to deflection by human intelligence or consciousness as gravitation itself. We *think* we are free to have fewer children than we actually have, in just the same way as we used to think, before the seventeenth century, that we were free to raise water more than thirty-three feet in a suction pump. But in reality we were not and are not free in either of those cases, or in any other case: merely ignorant of our chains. This is the conception of all life, and in particular of human life, which Darwinism attempts to establish as the true one.

This conception of life (as I have pointed out in earlier essays) is not true, because it is not true of human life. Despite Darwin—and despite Hume, Malthus, and Schopenhauer too—human life is not a plenum: it contains countless unrealized possibilities of reproduction. (See Essay 2.) Pine life may be a plenum, cod life likewise, but our life is not. In fact, on the contrary, in all civilized societies, the more opportunities for reproduction people have—that is, the more privileged they are—the *less* they reproduce. (See Essay 4.)

Of course a Darwinian, such as R. A. Fisher, will insist on calling this fact the "*inverted* birth rate" of all civilized societies.[65] But this is just a case of a characteristic vice of Darwinians, of *blaming the facts* for failing to agree with their theory.[66] It is also conjuring into existence, out of thin air, and merely in order to satisfy the demands of Darwinism, a Cave Man or Hobbesian "state of nature" in the past, when men were men, and the more privileged ones among them *were* the more prolific of offspring. Whether the same thing was in those days true of women is not revealed by Fisher, Huxley, Hobbes, or by any other expert on that interesting period of our history. (But then, there *are* no women in Hobbes's or Huxley's "state of nature," as anyone who has read those authors will recall: I suppose this was one of the things which made that state so eminently natural.)

Then, Darwinism requires the struggle for life in our species to be so severe as to exact a child mortality around 80 percent at all times. But human child mortality, during the only period for which we know anything at all about it, has hardly ever been near 80 percent, and during the last hundred years, in all countries for which reliable statistics exist, has hardly ever been a quarter of that figure. (See Essay 5.) We have therefore witnessed a direct disproof of Darwin's statement quoted above, that if you "mitigate the destruction ever so little . . . the number of the species will almost instantaneously increase to any amount." We *have* mitigated the destruction of human beings between birth and puberty, and mitigated it, not "ever so little," but enormously; and yet our numbers have *not* "almost instantaneously increased to any amount." (Our numbers *have* increased, of course, much more than some of us *like*. But that is an entirely different matter; and a matter of no relevance whatever to the truth or otherwise of Darwinism.)

Human life, then, straightforwardly contradicts the Darwinian conception of life at various points. But as well as that, human intelligence and consciousness plainly have a degree of autonomy which is wildly inconsistent with Darwinism. If intelligence and consciousness in humans are always subordinated, like all other adaptations of organisms, to their striving to increase, then *The Origin of Species* was an attempt by Darwin to increase the number of his descendants. But it was not. *Ergo*, etc. Similarly, some sociobiologists have realized that, according to the account which they themselves give of all communication—namely, that it is a form of manipulation by genes—*their own publications are simply power plays by their genes for increased representation in the next generation.* This realization is, very naturally, found both discouraging, and somewhat bewildering, by the authors concerned. For after all, when they *began* their careers, they had thought they were doing something entirely different: namely, biological science. Yet if what *The Selfish Gene* says is true, what else can that book be, but manipulation of its readers by the genes of Richard Dawkins, striving for their own maximal replication? Thus does the new religion, like revolutionary Marxism, consume its own devotees. But then, they had "asked for it," if ever anyone did.

In fact the autonomy of the human intellect will always present insuperable obstacles to *any* attempt at the complete mechanization of the picture of life. For suppose that some super-Darwin of the future attempts in a book to reduce every concept of biology, without remainder, to concepts of basic physics; and suppose he *succeeds in* this attempt, and is even recognized to have succeeded. This would certainly be a book of unparalleled scientific importance: an event that would put everything else in the history of thought absolutely in the shade. But how could this book *itself* be translated without remainder into the language of basic physics, while still saying what it said at first?

Darwinism, as I have implied, has always been governed by the idea that scientific progress involves recognizing as impossible things we had previously thought possible, and recognizing human freedom and power as being less than we previously thought. Well, it often does involve those things, in certain respects: perhaps always does. But then, scientific progress *also* often involves recognizing as possible things we had previously thought impossible, and recognizing as being *within* human power things we previously thought altogether beyond it. Everyone knows this, because all the "marvels of modern science and technology" are of this second kind, and tend not to the contraction, but to the enlargement, of human freedom and power. The physical discoveries of the seventeenth century, it is true, produced little if anything of this kind. What could anyone *do* in 1700, because of the progress of physics, which had been considered impossible in the year 1600? I venture to say, nothing at all; but certainly extremely little if anything. But this fact simply shows that seventeenth-century physics furnishes a very one-sided sample of scientific progress in general. Increased "mechanization of the world picture" is *not* the only, or even the invariable, outcome of scientific progress.

IN THE PRESENT ESSAY, however, my purpose is not to establish again that Darwinism is false: it is to explain how the new religion of genes came about.

But what I have said about the Darwinian conception of life has not been irrelevant to explaining the origin of the new religion. On the contrary, now that we have, clearly before our minds, the Darwinian conception of life, it is quite obvious how the new religion came about. Indeed, the question virtually answers itself. For that religion *is* just the old Darwinian conception, supplemented by certain details drawn from genetics.

Recall, first, that both old Darwinism and the new religion conceive life as through and through purposive. What drives the whole process of struggle, natural selection, and evolution, according to Darwin, is the striving of organisms to increase. And (as I said earlier in this essay) if one were to take out of the new religion every purposive disposition or activity which it ascribes to genes—selfishness, manipulation of various things, competition with one another, the use for their own ends of countless tactics and tools—then there would be nothing left of the new religion at all. New religionists, such as Williams, Dawkins, and Wilson, regard people and all other organisms as the helpless puppets, tools, or vehicles, of hidden purposive agents of more than human power and intelligence, whose only goal is to produce the largest possible number of their replicas in the next generation of organisms. But then, as we have now seen, Darwin, Schopenhauer, Malthus, and Hume regarded all organisms in essentially the same way: as mere means, employed by immensely powerful purposes, utterly foreign and unknown to the organisms themselves, aimed at producing the greatest possible number of descendants of the organisms.

The purpose which rules all organisms, Schopenhauer called "the Life Force," and Darwin called it "the striving to increase." These names, of course, are no longer current. But when new religionists say (as they all do say) that the organism is only DNA's way of making more DNA, that organisms and their adaptations are means which genes make and employ for their own ends (etc., etc.), then it is perfectly clear that the *thought* of the new religionists and that of the old Darwinians is one and the same.

According to Schopenhauer and Darwin, organisms are merely vehicles of an immensely powerful agency which, though purposive (and therefore intelligent at least to some degree), is as unconscious as gravitation or inertia, and as incapable as those forces of being deflected by anything that organisms know or intend; an agency which is distributively resident in the bodies of all organisms; which is bent only on "making more of the same," and as many more as possible; but which is constantly restrained from achieving unlimited increase, by the limitedness of food, and the struggle for life which results from that limitation. Now, will anyone deny that this is *also* the conception of life of Williams, Dawkins, and Wilson? So *far*, there is no difference at all.

The novel element, which *distinguishes* the new religion from the Old Darwinism, is of minor importance by comparison. It is the fact that Schopenhauer and Darwin, and indeed all Darwinians up to about 1930, thought only in terms of the whole organism, and in terms (so to speak) of "one packet of Life Force per customer"; whereas the new religionists have learnt better than that. *They* think of any one organism as being the puppet of many masters at once.

The reason is, of course, that they are among the intellectual beneficiaries of Mendel. Everyone must always have known that what we call "reproduction" never results, among sexual organisms, in an offspring which is in all respects identical with either of its parents. But no one knew, before Mendel brought the fact to light, that reproduction, in its fine grain, *does* consist of many independent strands of exact replication. Genes, and genes alone in the organic realm (leaving aside asexual organisms), really do "reproduce *themselves*" and nothing else.

It perhaps deserves emphasis (though I have already implied), that no old Darwinian, either before or after Darwin, ever thought of the hidden rulers of life as situated *outside* the individual organisms which they rule. Schopenhauer expressly says that the Life Force is in the body of each organism,[67] and Darwin, or any old Darwinian, would have taken that much for granted. Genetics has merely provided the new religionists with the *precise* locality of their gods, on the chromosomes of the sex cells.

The basic idea of the new religion, then, that humans and all other organisms are mere means to the ends of more powerful intelligent agents, is not an innovation of the last few decades. On the contrary, it was present all along, in the conception of life which Darwin shared with Schopenhauer and some others. The purposive gene gods of the new religion *are* the Life Force of Schopenhauer or the striving to increase of Darwin; only broken up into a multitude of little independent life forces or strivings to increase, in each single organism, and "given a local habitation" in its body. *That* is how the new religion came about.

Essay 11
Errors of Heredity, or
the Irrelevance of Darwinism to Human Life

. . . we may feel *sure* that *any* variation *in the least degree* injurious
would be *rigidly destroyed* [by natural selection].
—Darwin, *The Origin of Species*

D O YOU REALIZE, reader, that you are an error of heredity, a
biological error? Anyway you are, whether you realize it or
not. And not only *an* error, but an error on an enormous scale.
At least, Darwinians say you are. And who knows more about
biology and heredity, pray, than they do?

It does not sound a good thing to be, does it, a biological
error? In fact it sounds horribly like what you yourself, in mo-
ments of depression, have often suspected that you are. Well, it is
not a good thing to be. Even at the best of times, an error of
heredity has a distinctly short future, and the larger the scale
of the error, the shorter its future. An error on the scale that you
are has no future at all, to speak of. The one gleam of consola-
tion in all this is that just about every other human being, past or
present, was or is a biological error too, and on roughly the same
scale as you.

A biological error, or error of heredity, is an organism which
does not have as many descendants as it could have, or a charac-

teristic of an organism which prevents it having as many descendants as it otherwise could.

Among plants there is no biological error at all, and in most species of animals there is none worth mentioning. A cockroach, a fish, or a snake, hardly ever has fewer descendants than it could. They do not waste their time or their health on biology, or philosophy, or religion, or art, or social reform, or any such foolishness. They don't smoke, drink, or gamble either, nor yet do they practice contraception, or fret themselves about over-population or the environment. They concentrate all their efforts, from the earliest possible moment, on having as many descendants as they can. Nor do they often fall short of this goal, and even when they do, it is seldom by much.

In our species, by contrast, biological error, and on a large scale too, is absolutely rife. And the result is, of course, that hardly any members of our species ever have anything like as many descendants as they could. The following are a few examples of our errors which have been remarked upon by distinguished Darwinians in recent decades.

First, people who are naturally celibate, and never feel any strong interest in copulating with a member of the opposite sex. They are, of course, only a small minority of all people. But they are not so very uncommon that most of us have not encountered a few of them; and more than a few who seem to depart only slightly from being natural celibates.

Here is a famous Darwinian, C. D. Darlington, on the subject of the naturally celibate. "According to Galton's way of thinking, which all later study confirms, the natural celibate is an individual lying at the end of a curve of errors. He arises, as we may say, by a combination of errors of heredity."[1] That was, indeed, "Galton's way of thinking," but not only his: it was 100 years ago, and still is, the way of thinking of all Darwinians.

A second example of biological error in humans (though one which is not confined to humans) is pointed out by Professor E. O. Wilson. This is, an animal who, in fighting with a conspecific, when getting the upper hand, accepts signals of submission from its opponent. Among humans, as among dogs, such signals may be physiological or behavioral, and they usually have the effect of preventing a fight from ending in the death of the loser. But accepting submission signals is plainly a biological error, or, as Wilson puts it, constitutes for Darwinism "a considerable theoretical difficulty: why not always try to kill or maim the enemy outright?"[2] Why not, indeed? For to accept your enemy's signals of submission is to allow him to survive, probably to reproduce, and possibly even to fight you again another day, with more success.

A third biological error among humans (though among certain monkeys as well) is pointed out by Dr. R. Dawkins. This is, a bereaved mother's stealing and "adopting" another mother's baby, and the real mother's resenting this "baby snatching." As Dawkins says, there is here biological error on *both* sides. "The adopter not only wastes her own time: she also releases a rival female from the burden of child-rearing, and frees her to have another child more quickly. It seems to me a critical example which deserves some thorough research. We need to know how often it happens; what the average relatedness between adopter and child is; and what the attitude of the real mother of the child is—it is, after all, to her advantage that her child *should* be adopted: do mothers deliberately try to deceive naive young females into adopting their children?"[3]

But these four things—natural celibacy, accepting submission signals, and baby-snatching and the resentment of it—are only the beginning of the tale of our biological errors. We are guilty of many more errors which need no Darwinian experts to point them out, because they are perfectly obvious even to mere street Darwinians like ourselves. One of these is our proneness to passions which are not merely not conducive to reproduction, but positively inimical to it.

One such passion is the love of truth. For it is always leading some people—admittedly never more than a small minority—to devote their lives principally to science, or to history, or to philosophy. Another such passion is the love of beauty. Think, for example, of the love of beauty in music. Have not millions of members of our species devoted a great deal of their time to listening to Bach, or Rameau, or Mozart, or whoever: time which they could perfectly well have devoted instead to ensuring that they had more grandchildren? In fact this error is so very common and glaring that it is useless to try to deny it.

But if the love of beauty in music is an inveterate error with many of our conspecifics, what shall we say of homosexuality? It is of immemorial antiquity, widespread, and to all appearances absolutely incorrigible. Yet what biological error could possibly be more glaring than homosexuality? It probably ought to be classed under the old Catholic heading of *invincible* error.

Then there is another biological error, as ancient as homosexuality, but even more widespread and inveterate in our species, and absolutely peculiar to our species. I mean, the practices of preventing conception, of abortion or feticide, and of infanticide. As Darwin says, "the instincts of the lower animals are never so perverted as to lead them regularly to destroy their own offspring . . . ."[4] But we are different, and have been so from as far back as historical knowledge extends.

Contraception, homosexuality, natural celibacy, the love of truth or of beauty, accepting submission signals, adopting children, and resenting baby snatchers: what a heavy catalogue of errors! It singles out our species as being the most hopelessly stupid of all the pupils in the great school of natural selection. For a species of insects or fish or snakes that fell into any one of those errors to a significant extent would soon have its mistake corrected, with the utmost severity, by natural selection. Or at least, that is what Darwinism says. So how come *we* are allowed to get away with committing all these errors at once? How have we managed to survive at all under all these multiplied handicaps?

And yet the errors of heredity which I have so far mentioned are *still* only a fraction of all those to which our species is a prey. In fact our errors are so many that, simply to keep my thoughts for the present essay in some sort of order, I have had to compile an alphabetical list of them. My list is certain to be very incomplete, but it is already much too long to be reproduced here. It will suffice, however, to give the reader some idea of its length, if I give a selection of some of the items under just two letters of the alphabet.

Under the letter A, the entries include:

Abortion
Adoption
The Popularity of Alcohol
Altruism
Love of, and from, Animals of other species
The importance attached to Art
Asceticism (sexual, dietary, or whatever)

The entries under H include:

Heirs who respect the wishes of dead parents
Heroes
The admiration of Heroism
Homosexuality
The idea of Honor
Horror at the struggle for life in other species
Humaneness
Humor

Nearly all of these things are peculiar to our species, and all of them are plainly prejudicial to having as many descendants as one could. How can you help your chances of reproduction, or do anything but injure those chances, by having as your best mate a dog, a horse, or a cat? Or by being afflicted with a sense of honor? Or by worrying about the wishes of certain other people who are no longer even alive?

That heroism is an error of heredity will be obvious. A hero, by the very meaning of the word, is more likely than a non-hero to have his reproductive career cut short at an early age. But the admiration of heroes is a biological error too, though a less obvious one. It can clearly do us harm by leading us to imitate heroes, and thus to increase the probability of our dying young. But how could it possibly do us any *good*, as Darwinians understand "good"? That is, how could it possibly help us to have more descendants? And yet this admiration, though injurious or useless to those who feel it, is and always has been extremely common: in fact, almost universal. From Homer's time to ours, admirers of heroes have always been much commoner than heroes. Why hasn't this admiration been "selected out"?

These twin errors, of heroism and the admiration of it, have a special historical interest. For they were discussed at length by R. A. Fisher in *The Genetical Theory of Natural Selection* (1930).[5] That is, by the most penetrating intelligence, and in the most seminal book, of all twentieth-century Darwinism. But Fisher found both heroism and the admiration of it just as inexplicable as run-of-the-mill Darwinians find them. Even to explain the former, he is obliged to postulate—on no evidence at all, and in defiance of plain probability—that in "barbarian" societies a hero has *more* descendants than a non-hero. As to explaining the admiration of heroism, he can scarcely be said even to have attempted it.

At one point in his discussion Fisher writes as follows.

> The hero is one fitted constitutionally to encounter danger; he therefore exercises a certain inevitable authority in hazardous enterprises, for men will only readily follow one who gives them some hope of success. Hazardous enterprises, however, are not a necessity save for the men who, as enemies or leaders, make them so, and the high esteem in which tradition surrounds certain forms of definite imprudence cannot be ascribed to any just appreciation of the chances of success.[6]

This is one of those passages in which a writer unwittingly reveals that he is utterly unable to enter into the minds of the people he is writing about. In Fisher's eyes, we suddenly realize, the world that humans live in is so perfectly safe, that there is *never* any *necessity* for hazardous enterprises! They are only *made* necessary by the restless nuisances called heroes; in order (one supposes) to exercise their peculiar talent. The point of view of Sancho Panza, or of suburban man, could hardly have been expressed more ingenuously.

But the Sancho Panza or suburban point of view is also (as has often been pointed out) that of Darwinism: the point of view of the prudent organism, unsleepingly attentive to its own survival and reproduction, and careless of all else. And it must be admitted that, if maximizing the number of one's descendants is the goal of all life, then the winners of the Victoria Cross (for example) have each been guilty of "certain forms of definite imprudence." "Oh God! Oh Montreal!," (as Samuel Butler remarked in a similar context).

IT IS EASY to see how Darwinians came to think of natural celibacy, contraception, and all the other things I have mentioned as *errors*. For they know that, almost without exception, organisms act as though in obedience to one supreme imperative: namely, "Make some more things of the same kind as yourself, and as many of them as you can." Everything about a pine, a fish, or a rabbit, and everything that they do, is subordinated to the goal of maximal survival and reproduction. And since, in virtually all species, an organism can reproduce only while it does survive, this goal is better expressed as simply that of maximal reproduction. So then, when Darwinians turn their attention to our species, and find there, though nowhere else, contraception and lots of other things which have an anti-reproductive effect, they understandably form an impression that *something has gone wrong* with this species.

It is rather as though soccer were now, and always had been, the only game anyone ever played, or ever heard of, and everything else in human life were subordinated to the goal of winning at soccer; and then one day we came across a solitary pair who were playing a game in which it was part of the object to make a ball go *over* a net. In these circumstances it would be natural enough, would it not, if we thought that there must be something wrong with these two people, or that they had made some sort of mistake?

Natural enough, perhaps; but completely irrational. What evidence is there that there is *anything* wrong with these people, intellectually, physically, morally, or in any way? They are doing something uncommon, certainly: everyone else obeys the imperative "Try to win at soccer," and they do not. But that is absolutely all there is to it. It might of course happen to be *true* that these two people have made a mistake. They might be trying to play soccer, and just be so exceedingly stupid that playing tennis is the closest they can get to it. But the rational thing to believe is that they have not made any mistake, but are simply not trying to win at soccer.

It is far worse than irrational, however, to call natural celibates, adoptive parents, heroes, and so on, "errors." It is inexcusable. There are two reasons.

One of them is that it does not make sense. What is an error? A falsity that is taken for a truth, of course. But an actual organism, or a characteristic of an actual organism, cannot possibly be a falsity. That simply makes no sense. Propositions, beliefs, utterances, etc., have (as logicians put it) truth values, and falsity is one such value. But people, or organisms of any kind, do not have truth values. A man who is naturally celibate, or the practice of contraception by a woman: these are simply not things of a kind that can intelligibly be described as either false or true. They can no more be true or false than they can be even or odd.

The second reason why it is inexcusable to call organisms, or characteristics of them "errors" is even more important. It is this: that you cannot call something an error without *reprehending* it in some way. An error—there is just no getting around this—is a bad thing: it is getting something, or doing something, wrong. It is no good saying (as many people nowadays would like to say about, for example, many of the beliefs of Australian aboriginals), "it's an error but, mind you, I'm not criticizing it."

That is at best a joke, and otherwise is just a self-stultifying speech act: like saying (for example) "I promise to repay my debt but, mind you, I'm keeping *all* my options open." By the time you have called something an error, it is too late to claim that you are not reprehending it. You already *have* reprehended it.

An inevitable consequence is that Darwinians, whenever they identify something as being a biological error, cannot avoid reprehending or condemning or criticizing it in some way. They do not have actually to *call* the thing a biological error, of course. They have plenty of other ways of letting us know which things they do regard as biological errors. And they *have* let us know (as we have seen) that they do so regard abortion, adoption, alcoholic drinks, and the love of animals—to remind you of a few of the many errors to which our species is peculiarly prone.

The Darwinian reprehension of biological errors comes in a range of colors (so to speak). Sometimes it is principally intellectual: that is, it is reprehension of the *stupidity* of the organisms in question, which cannot see where their interests in survival and reproduction lie. But this shades into prudential reprehension: of the *folly* of organisms which, though they can see, do not regularly act on, their own interests. And that in turn, at least where human biological errors are concerned, sometimes shades into *moral* reprehension. But reprehension of some kind and shade there must be, as soon as Darwinians do identify something as a biological error.

We have already encountered some examples. Darwin, in the passage quoted above, clearly expressed a *moral* reprehension of the peculiarly human errors of infanticide and feticide. He elsewhere expressed (as I pointed out in Essay I above) moral reprehension of the related biological error of contraception; though there was probably in that case a tincture of prudential reprehension as well. Darlington, in the passage quoted above about natural celibates, expressed plainly enough his own and Galton's reprehension of that class of persons; though it would not be easy to decide how far this was moral reprehension, and how far prudential. Dawkins, in the passage I quoted about baby-snatching, implied as plainly as anyone could, that mothers are dishonest if they only seem to resent baby snatching, and stupid if they really do resent it. Fisher, in the passage I quoted about hazardous enterprises, unmistakably implied a criticism of heroes: a criticism which (since the *imprudence* of heroism goes without saying) I take to have been a moral one. A more out of the way example is the Darwinian reprehension of alcoholic drinks. Alcohol was one of the countless bugbears of Charles's grandfather, Erasmus Darwin. He died in 1802, but he managed to pass on his reprehension of alcohol to at least the fourth generation of his Darwin descendants: that is, to Charles's grandchildren.[7] In this case the reprehension was certainly principally prudential; but it always had in addition a strong infusion—perhaps 40 percent—of moral reprehension.

There was another biological error which Erasmus Darwin especially reprehended: clerical celibacy. And this reprehension, too, he managed to pass down a long way: at least as far as Charles Darwin's children. Darwinians nowadays, of course, all hate the Catholic church because of its opposition to most forms of contraception. But throughout the nineteenth century they all hated it for an almost opposite reason: the celibacy which it imposes on its clergy. In Darwin households, and closely related ones, no two opinions were possible about clerical celibacy: it was obligatory to be brim full of indignation about it. Charles's eldest daughter, Henrietta Litchfield, got an especially heavy dose of this hereditary indignation; from which it is a reasonable inference that Charles himself had got an especially heavy one. His cousin, Francis Galton, quite certainly did.

The reason that Galton gave in print, for reprehending clerical celibacy, was this. That by it, European civilization had been willfully deprived, for nearly 1,000 years, of all the valuable progeny which could have been expected from many of the best and ablest men and women who ever lived. Think of the intellectual endowments of an Aquinas, and of the moral gifts of a St. Francis or a St. Teresa of Avila. And then recall that, by the deliberate policy of the Catholic church, all of this vast fund of heritable human wealth has been condemned to perish with the bodies of those celebrated individuals.[8]

Galton was the founder of the eugenics movement, and his reasoning about clerical celibacy naturally recommended itself to all his fellow eugenists, who were, of course, Darwinians without exception. For my own part, however, I do not believe that his published reason was the only or even the main reason for Galton's detestation of clerical celibacy. In fact I think it was a smokescreen.

For a start, it is *childlessness* which is the unforgivable sin against the holy spirit of evolution: childlessness as such, no matter whether it be voluntary or involuntary, or how it may be brought about. It is nature's law, and the goal of all life, to have as many descendants as possible. What reprehension is too severe, then, for those who have none at all? And yet Francis Galton, Henrietta Litchfield, and Darwin's most devoutly eugenist son, Leonard (who was also the mentor of the eugenist R. A. Fisher), were all, though married, childless themselves. What an infinity of shame they must all have felt! No wonder Leonard Darwin labored all his life under a strong conviction of inferiority, not only to his father, who had seven children, but to his four brothers, who at least managed to produce nine children among them.

But even if we suppose that it was exclusively (though illogically) *voluntary* childlessness which excited the reprehension of Darwinians, why did they confine their indignation to *clergy* who were voluntarily childless? There are plenty of other classes of people, after all, who are voluntarily childless.

One such class is the men and women, common enough in every age, who are childless only because of their unflagging practice of contraception or feticide. Now, that practice requires intelligence, diligence, and self-control: good qualities which would be of great benefit to society as a whole if only they were allowed to descend to offspring. And then, there are all the women who decline every offer of marriage in order to devote themselves to caring for an aged parent, an invalid relative, or the children of a bereaved brother-in-law. These people, too, deliberately deprive society of the offspring who would have inherited, at least to some extent, their uncommon nobility of character. Why, then, were the Darwinians not just as indignant about *them*, as they were about Catholic priests and nuns?

Darwin's daughter Henrietta was not a writer, so we do not know what reasons she would have given in print for her furious indignation about clerical celibacy. But she appears, from a wonderful account given of her by one of her nieces, to have thought that clerical celibacy could only be an invention of cruel spoilsports.[9] If it was so, then her reprehension of clerical celibacy was simply a late and partial outcrop of the *old* evolutionism, of the late eighteenth and early nineteenth centuries, to which I have referred in Essays 2 and 10 above: the evolutionism of the time of her great-grandfather Erasmus. In those revolutionary days, evolutionism had been a package deal which had universal sexual emancipation, and good times had by all, as a prominent component, along with regicide, anti-clericalism, and atheist terrorism. This was, of course, the evolutionism which Charles Darwin labored all his life to live down; partly by the simple stratagem of saying as little as humanly possible about his grandfather Erasmus who, in addition to all his other offences, had inconsiderately anticipated so very much of the contents of *The Origin of Species*.

This was also the real ground, I believe, of the detestation of clerical celibacy among nineteenth-century Darwinians in general. It was the old anti-clericalism, and sexual emancipation, coming to the surface in the uncongenial environment of late nineteenth-century England. If so, then Galton's lament about the wonderful children that Aquinas or St. Francis might have had was as hollow as it always sounded. His hypothesis at least explains (as I say) why it was *clerical* childlessness alone that evoked Darwinian reprehension.

As an example of the Darwinian reprehension of biological errors, indignation about clerical celibacy may be a little out of date. But everyone nowadays is perfectly familiar with this reprehension, even if they have never read a single word of Darwinian literature. The reason is that there is now so much of it about. A perfect example of it is the reprehension of smoking by our health police. I take this to be about 15 percent intellectual, 25 percent prudential, and 60 percent moral; but even if these percentages are out, that is certainly the right ordering in size of the three components. The health police are, of course, altogether an invention of Darwinism; as Samuel Butler, more than a hundred years ago, foresaw that they would be.[10]

BUT NOW LET US step back for a minute, stop thinking like Darwinians, and try to think like people of sense instead. Then we will recall two very important things.

One of them is that scientific theory cannot possibly reprehend, in any way at all, any actual facts. It can explain them, predict them, describe them, but it cannot condemn them as errors. Astronomy cannot criticize certain arrangements of stars or planets as erroneous, and no more can biology criticize certain organisms, or characteristics of them, as erroneous. Those organisms and their characteristics are simply among the realities which biology exists to describe and explain. As moralists we can find these characteristics admirable or the reverse, and as lovers of beauty we can find the organisms glorious or hideous. But as scientists we cannot do any of those things, or anything like them.

The other important thing that will now come back to us is that there is *nothing at all wrong with* being naturally celibate, with a man's accepting submission signals in a fight, with a mother's resenting baby snatchers, and so on. Speaking essentially, that is. No doubt *in fact* every one of the people who have those characteristics has got any number of things wrong with them: medically, morally, or intellectually. But that is only (as philosophers say) *per accidens*. A person is not reprehensible *in virtue of* being naturally celibate, of accepting submission signals, of resenting baby snatchers, of heroism, of being a lover of horses, etc.

Even those who think that feticide is morally reprehensible do not think it so *because* it reduces the number of descendants someone has. They would still equally reprehend it, even where it did not in fact have that effect: even where (for example) a woman, every time she arranged for a fetus of hers to be aborted, also arranged for another one of her fertilized eggs to be implanted in another woman, there to be brought to birth and a normal span of life.

WITH THESE TWO wholesome recollections present to our minds, let us now return to Darwinism, and demand to know what the hell is going on, with this business of calling certain things "biological errors," "errors of heredity," and the like?

The answer is simplicity itself: what is going on is just this. Wherever *Darwinism* is in error, Darwinians simply call the organisms in question or their characteristics, an error! Wherever there is manifestly something *wrong with their theory*, they say that there is something wrong with the organisms. Their theory implies that there is *no such thing* as natural celibacy, contraception, or feticide, and where all other species are concerned, it is true that there is no such thing. But in our species, those and many other anti-reproductive characteristics do exist. And so Darwinians, rather than admit that their theory is simply not true of our species, brazenly shift the blame, and designate all of those characteristics "biological errors."

The moral arrogance of this proceeding should be sufficiently obvious. Recall the contempt, tinged with pity, which Galton and Darlington felt for the naturally celibate. The class of natural celibates happens to include, among others, such contemptibles as Gibbon and Macaulay, Vivaldi and Handel, Locke, Leibniz, Newton, Kant and Mill. Now, was either Galton or Darlington, do you think, entitled to feel contempt of *any* sort for *those* people? Is Richard Dawkins entitled to imply that every mother who seems to resent baby snatching is either dishonest or stupid? Is E. O. Wilson entitled to wonder why he does not try to kill or maim certain other Harvard biology professors who are, notoriously, his bitter enemies? Was Fisher entitled to make a moral criticism of *heroes*? Talk about moral arrogance—"nerves of steel" would be a better word for it!

But the moral arrogance of Darwinians is thrown entirely into the shade by their intellectual arrogance. Because what their theory says about man is badly wrong, *they* say that *man* is badly wrong: that he incorporates many and grievous biological errors. What a great idea: it saves so much trouble!

When Isaac Newton was first trying out the theory that gravitational attraction obeys an inverse square law, he happened to make use of some erroneous data about the radius of the earth, and as a result various things, including the moon's distance, came out badly wrong. He promptly lost confidence in the theory, and put it aside for many years. Now why didn't he think of doing the *Darwinian* thing, and simply say that the *moon's distance is wrong*: an astronomical error? I suppose this must have been just another one of his biological errors, to go with his natural celibacy.

Can intellectual arrogance—not to say madness—go further than this Darwinian business, of categorizing many human attributes as errors? Why yes it can, and among Darwinians it often does. For while they usually imply that man is guilty of many and grave biological errors, they sometimes go the other way. That is, they sometimes, when at the height of intoxication with their theory, imply that there really is *no* biological error, or only maximally evanescent error, anywhere: not even in man. In other words, they imply that we *do* obey, just as infallibly as pines and flies and fish and rabbits do, the great biological imperative, to maximize the number of our descendants.

Here, for example, is a respected sociobiologist, Professor R. D. Alexander, writing in 1979: ". . . we are programed to use all our effort, and in fact to use our lives, in reproduction."[11]

This is one of those statements which are so breathtakingly false, that initially their only effect on the reader or hearer is to produce stupefaction. One can at best only gasp out something like, "Well in that case the program isn't working, and never has worked." People who use all their effort, in fact use their lives, in reproduction: does that sound like anyone *you* know, or ever heard or read of? I cheerfully grant to Professor Alexander, that every fly, fish, or rabbit does have as many descendants as it can. But will he be able to point out to me, in return, even *one* man or woman who does the same, or ever did so? And yet he *says* that *all* of us do!

You might think that a falsity so astounding as this statement of Alexander must be an isolated aberration. But if you do think so, then you know little indeed of the literature of Darwinism. Several other contemporary Darwinian authorities could easily be quoted to exactly the same effect. In fact this ultimate degree of Darwinian faith, which blinds the faithful to even the most obvious facts of human life, is far commoner at present than it ever was before. But rather than quote other contemporary examples of this blindness, it will be more useful if I point out that Alexander has on his side the very best of Darwinian authorities: Darwin himself.

The actual phrases "biological error," "error of heredity," and the like are not of very common occurrence in Darwinian literature. But that is no more than an accident of expression, because the *idea* of biological error is omnipresent in that literature. Darwin's usual phrase for it was "an attribute which is injurious to its possessors in the struggle for life"; or more shortly, "an injurious attribute," or a "disadvantageous" attribute. An injurious or disadvantageous attribute is simply one which lessens an organism's chances of surviving and reproducing. So the things we have been calling biological errors—feticide, accepting submission signals, and all the rest—are still our subject. It is just that I will now usually call them, as Darwin usually did, "injurious attributes."

The first chapter of *The Origin of Species* is about domestic animals. Darwin proves, by many examples, that new varieties of domestic species have often been brought into existence by breeders or fanciers. A certain new characteristic crops up in a few members of a herd of flock, and if this attribute is useful or otherwise interesting to the breeders of fanciers, they seize upon it, and breed selectively for it. If they succeed, the result is a new variety, brought about by "artificial selection."

The second chapter is about how much variation there is in natural populations, and the third is about the struggle for life in nature. Then Darwin returns to the subject of new kinds of organism being produced by selection from existing kinds, and he begins Chapter IV, "Natural Selection," as follows.

> How will the struggle for existence, discussed too briefly in the last chapter, act in regard to variation? Can the principle of selection, which we have seen is so potent in the hands of man, apply in nature? I think we shall see that it can act most effectively. Let it be borne in mind in what an endless number of strange peculiarities our domestic productions, and, in a lesser degree, those under nature, vary; and how strong the hereditary tendency is. Under domestication, it may be truly said that the whole organisation becomes in some degree plastic. Let it be borne in mind how infinitely complex and close-fitting are the mutual relations of all organic beings to each other and to their physical conditions of life. Can it, then, be thought improbable, seeing that variations useful to man have undoubtedly occurred, that other variations useful in some way to each being in the great and complex battle of life, should sometimes occur in the course of thousands of generations? If such do occur, can we doubt (remembering that many more individuals are born than can possibly survive) that individuals having any advantage, however slight, over others, would have the best chance of surviving and of procreating their kind? On the other hand, we may feel sure that any variation in the least degree injurious would be rigidly destroyed. This preservation of favorable variations and the rejection of injurious variations, I call Natural Selection."[12]

His theory, of the struggle for life and of natural selection, was not put forward, of course, just for some species but not others. Darwin put it forward for all species whatever. The passage just quoted is therefore about our species, quite as much as it is about any other. And now consider part of what it says. Namely, that we may feel sure that any attribute *in the least degree* injurious would be *rigidly destroyed*.

Think what that means. It means that an impediment, however microscopic, to an organism's maximal reproduction, if such an impediment ever occurred at all, would be completely eliminated by natural selection, and as quickly as such elimination can take place. In other words, biological error is either non-existent or is maximally evanescent. Or, in other words again, the members of our species, like the members of every other, do have as many descendants as they can. Which is, after all, just what Alexander said; and has the same stupefyingly obvious falsity as the statement which I quoted from him.

For there are in our species many attributes which are injurious to their possessors, not only "in the least degree" but *extremely* injurious, and which are nevertheless not "rigidly destroyed." Some of them, such as accepting submission signals, adoption, and the maternal resentment of baby snatching, since we share them with a few other species, are presumably in fact *older than* our species. Others, such as feticide, contraception, homosexuality, the love of animals, fondness for alcoholic drinks, and respect for the wishes of the dead, extend back to the remotest antiquity of which we know anything, and far from being "rigidly destroyed" show no sign of diminution.

Of course there are many other characteristics, besides those I have so far mentioned, which are peculiar to our species, persistent in it, and yet extremely injurious to reproductive success. One of the most obvious, and therefore one of the oftenest noticed, is the unparalleled dangerousness in our species of parturition, to both mother and child. Another obvious one is our ancient propensity for committing suicide. (Although, since there already is at least one sociobiologist who thinks that homosexuality is a device for enhancing your reproductive success—your *inclusive* reproductive success, of course—no doubt some sociobiologists think the same about suicide.)[13] A third injurious attribute which persists in our species—and a biological error which I greatly regret having been guilty of myself—is the reading of hundreds of books about evolution, which abound with idiotic statements such as the ones I have just quoted from Alexander and Darwin.

In fact, far from every attribute being rigidly destroyed which is in the least degree injurious, in *our* species there is precious little *except* injurious attributes. Nearly everything about us, or at least nearly everything which distinguishes us from flies, fish, or rodents—all the way from practicing Abortion to studying Zoology—puts some impediment or other in the way of our having as many descendants as we could. From the point of view of Darwinism, just as from the point of view of Calvinism, *there is no good in us*, or none worth mentioning. We are a mere festering mass of biological errors.

Which means, of course—once you turn that statement the right way up—that on the subject of our species, *Darwinism* is a mere festering mass of errors: and of errors in the plain honest sense of that word too, namely, falsities taken for truths. Darwinism can tell you lots of truths about plants, flies, fish, etc., and interesting truths too, to the people who are interested in those things. But the case is altogether different, indeed reversed, where our own species is in question. If it is *human* life that you would most like to know about and to understand, then a very good library can be begun by leaving out Darwinism, from 1859 to the present hour.

If you want to learn something about *Heroes and Hero-Worship*, for example, then you can do so by reading Carlyle's little book of that title: but it is a perfect waste of time to read on this subject what was written by that doyen of twentieth-century Darwinians, R. A. Fisher. Well, it could hardly have been otherwise. Would you expect to learn anything about the mind of a Pascal or a Schopenhauer, a Shakespeare or a Mozart, by reading what their respective bank managers wrote about them? For Darwinians, even according to their own account, are only *investment advisers* after all. They happen to occupy themselves with "the currency of offspring" (as Fisher taught them all to say), rather than with the currency of cash, that's all.

325

Gwen Raverat was a daughter of Charles Darwin's son George. She wrote a wonderful book entitled *Period Piece* (1952) about her childhood and her numerous Darwin relatives. Late in that book she remarks that the Darwins in general "were quite unable to understand the minds of the poor, the wicked, or the religious."[14]

This is most profoundly true. And it is true not only of Darwins, or of Darwinians of the blood royal such as Galton, but of all Darwinians of what might be called "the pure strain" of intellectual descent from Darwin: for example Fisher, Darlington, E. O. Wilson, and Richard Dawkins. And it means, of course, a rather large gap in their understanding of human life; since the poor, the wicked, and the religious, must make up, on any estimate, at least three-quarters of all human beings.

But true as Gwen Raverat's remark is, and far as it goes, it does not go nearly far enough. For there are many and large classes of people who are neither poor nor wicked nor religious, but who are still a closed book to the characteristically Darwinian cast of mind. They are the heroes, the adoptive parents, the men who do not kill every enemy they successfully fight, the intelligent mothers who detest kidnappers . . . but you know only too well by now how long the catalogue of our errors goes.

Acknowledgments

I owe thanks to far more people than I can name, for their critical comments on draft-parts of this book. But I owe a unique debt of this kind to my friend and former student Dr. James Franklin of the University of New South Wales. At every step of the way, his comments have been the most penetrating, his assent the most valued, and his dissent the most feared, of all those I have received.

D. C. Stove
October 1993

I must echo David Stove's thanks to Jim Franklin, Stove's friend and literary executor. Dr. Franklin, who arranged for the publication of *Darwinian Fairytales* shortly after Stove's death, has been enormously helpful as I prepared this new edition of the book. Readers of the previous edition, published in 1995 by Ashgate Publishing, will note that I have followed American conventions in spelling and punctuation, corrected a handful of typographical errors, and aggregated the citations at the end of the book.

Roger Kimball
December 2005

Notes

Essay 1 Darwinism's Dilemma

1   All the quotations in this paragraph are from T. H. Huxley, *Evolution and Ethics, and Other Essays* (London: Macmillan, 1894), pages 204–5, or, in the case of the single words or phrases, from the earlier pages of the same essay.
2   *Evolution and Ethics*, page 40, pages 210–2.
3   *Evolution and Ethics*, pages 40–1.
4   *Evolution and Ethics*, page 38. The italics are not in the text.
5   Charles Darwin, *The Descent of Man, and Selection in Relation to Sex*, 2d edition (London: John Murray, 1874), Vol. I, pages 205–6.
6   *The Descent of Man*, Vol. II, pages 438–9.
7   Mary Midgley, *Evolution as a Religion* (London and New York: Methuen, 1985), page 119.
8   See J. M. Robertson, *A History of Freethought in the Nineteenth Century* (Dawsons of Pall Mall, 1969), page 337, where Robertson is reporting a letter from Darwin to Annie Besant.

Essay 2 Where Darwin First Went Wrong About Man

1   Charles Darwin, *Autobiography* in *The Life and Letters of Charles Darwin*, edited by Francis Darwin (London: John Murray, 1888), Vol. I, page 83.

2   This was in a paragraph which occurs only in the second edition of
     Malthus's *Essay* (1803). The paragraph gave great and widespread
     offense, even more than the book as a whole did. It is quoted in full in
     Patricia James, *Population Malthus: His Life and Times* (London:
     Routledge and Kegan Paul, 1979), page 100, from which it is reproduced
     here.

> A man who is born into a world already possessed, if he cannot get
> subsistence from his parents on whom he has a just demand, and if the
> society do not want his labor, has no claim of *right* to the smallest
> portion of food, and, in fact, has no business to be where he is. At
> nature's mighty feast there is no vacant cover for him. She tells him to
> be gone, and will quickly execute her own orders, if he does not work
> upon the compassion of some of her guests. If these guests get up and
> make room for him, other intruders immediately appear demanding
> the same favor. The report of a provision for all that come, fills the
> hall with numerous claimants. The order and harmony of the feast is
> disturbed, the plenty that before reigned is changed into scarcity; and
> the happiness of the guests is destroyed by the spectacle of misery and
> dependence in every part of the hall, and by the clamorous
> importunity of those, who are justly enraged at not finding the
> provision which they had been taught to expect. The guests learn too
> late their error, in counteracting those strict orders to all intruders,
> issued by the great mistress of the feast, who, wishing that all her guests
> should have plenty, and knowing that she could not provide for
> unlimited numbers, humanely refused to admit fresh comers when her
> table was already full.

3   Charles Darwin, *The Descent of Man, and Selection in Relation to Sex*,
     second edition (London: John Murray, 1874), Vol. I, page 66, footnote.

Essay 3 "But What About War, Pestilence, and All That?"

1   T. R. Malthus, *An Essay on the Principle of Population* (London:
     Macmillan, facsimile edition of 1st edition, 1798), pages 43–4. Italics not
     in original.
2   Ibid., page 140.
3   Ibid., page 55.
4   Charles Darwin, *The Descent of Man, and Selection in Relation to Sex*,
     2d edition (London: John Murray, 1874), Vol. I, page 66. Italics not in
     original.
5   Ibid.
6   A. R. Wallace, *Natural Selection and Tropical Nature* (London and New
     York: Macmillan, 1891), page 25. Italics not in text. This book was a

republication of *Contributions to the Theory of Natural Selection* (1870) and of *Tropical Nature and Other Essays* (1878), with additions to and omissions from both these books. The passage I have quoted was therefore almost certainly first published in 1870.

Essay 4 Population, Privilege, and Malthus' Retreat

1   Thomas R. Malthus, *Essay on Population* (London: Everyman, 1927), Vol. II, page 261. (This edition is stated, on page xvi, to have been "reprinted from the seventh edition," with several omissions.)
2   Ibid., Vol. II, page 161.
3   Ibid., Vol. I, page 141.
4   Ibid., Vol. I, page 113.
5   Charles Darwin, *The Descent of Man, and Selection in Relation to Sex*, 2d edition, (London: John Murray, 1874), Vol. I, pages 69–70.
6   Thomas R. Malthus, *Essay on Population* (London: Macmillan, 1966) (facsimile reprint of the first edition of 1798), page 44.
7   See Darwin, *The Descent of Man*, page 69.
8   R. A. Fisher, *The Genetical Theory of Natural Selection* (New York: Dover Publications, 1958), page 252. And see the following pages to page 274.
9   Ibid., Chapter XII.
10   Ibid., pages 247–62.
11   In *Macmillan's Magazine*, August 1865.
12   Francis Galton, *Hereditary Genius*, edited by C. D. Darlington (London: Macmillan, 1892), page 187.
13   In *Fraser's Magazine*, September 1868.
14   One of them is given by Darwin, *The Descent of Man*, page 195, as being in *Anthropological Review*, May 1864.
15   William Godwin, *Of Population* (New York: Augustus M. Kelly, 1964), page 96. Italics not in text.
16   Malthus, *Essay on Population*, Vol. I, page 14.
17   Ibid., page 315. With respect to moral restraint, the editions of the *Essay* beginning with the second do not differ in any important respect.
18   See pages 363 and 368 of William Godwin, *Reply to Parr*, reprinted in *Uncollected Writings (1785–1822)* (Gainesville, Florida: Scholars' Facsimiles and Reprints, 1968), edited by Marken and Pollin.
19   Ibid., page 364.

Essay 5 A Horse in the Bathroom

1   A. R. Wallace, *Darwinism*, 2d edition (London: Macmillan, 1889), page 14.

2 Charles Darwin, *The Origin of Species* (Cambridge: Harvard University Press, 1966, a facsimile of the first edition of 1859), page 62.
3 Ibid., page 64.
4 T. R. Huxley, *Evolution and Ethics, and other essays* (London: Macmillan, 1894), pages 204–5. The single words or phrases are from earlier pages of the same essay.
5 Darwin, *The Origin of Species*, page 69.
6 Ibid., page 79.
7 Ibid., page 66.
8 Ibid., page 5.
9 Ibid., page 61.
10 Ibid., page 64.
11 See ibid., pages 65–6.
12 Ibid., page 64.
13 Ibid., page 66.
14 Thomas R. Malthus, *Essay on Population* (London: Everyman, 1927), Vol. I., pages 195–6.
15 Ibid., Vol. II, page 21.
16 Ibid., Vol. II, page 30.
17 Ibid., Vol. II, page 20. Italics not in text. Cf. Note 2 to Essay 2 above.
18 David Hume, *Essays Moral, Political and Literary*, revised edition, edited by Eugene F. Miller (Indianapolis: Liberty Fund, 1985), pages 452ff.
19 Richard Whately, *Introductory Lectures on Political Economy*, 3d edition (New York: A. M. Kelley, 1966), pages 231–2. The italics are in the text. I am indebted to Professor Antony Flew (in his *Thinking About Social Thinking* (London: Fontana Press, 1991), page 121, for this invaluable reference.
20 William Hazlitt, *The Spirit of the Age* (Oxford: Oxford University Press, 1945), page 151.
21 Namely, page 45, page 63, page 65 three times, pages 78–9, page 109, page 127, page 470. (All of these references are, again, to the first edition of 1859.)
22 Namely, page 5 and page 63.
23 Wallace, *Darwinism*, page 25.
24 See Richard Hofstadter, *Social Darwinism in American Thought* (Boston: The Beacon Press, 1959).
25 I have borrowed this quotation from Mary Midgley, *Evolution as a Religion* (London: Methuen, 1985), page 119; but she drew it from *Hitler's Table-Talk*, edited by Hugh Trevor-Roper (London: Weidenfeld and Nicolson, 1963).
26 W. T. Mills, *The Struggle for Existence*, 7th edition (Chicago and Manchester: International School of Social Economy, n.d. but about 1904).
27 S. Stepniak, *Nihilism As It Is* (London: T. Fisher, Unwin: n.d. but about 1893).

28   August Bebel, *Woman Under Socialism* (New York: New York Labor News Press, 1904).

29   Charles Darwin, *The Descent of Man, and Selection in Relation to Sex*, 2d edition (London: John Murray, 1874), Vol. I, page 399.

30   Ibid., Vol. I, page 67.

31   Konrad Lorenz, *Studies in Animal and Human Behaviour* (London: Methuen, 1971), Vol. II, pages 164ff.

32   Charles Darwin, *Vogage of H.M.S. "Beagle" Round the World* (London: John Murray, 1897), page 195.

33   See E. L. Bridges, *Uttermost Part of the Earth* (London: Hodder and Stoughton, 1951).

34   Lorenz, *Studies in Animal and Human Behaviour.*

35   For the statements in this paragraph, see Bridges, *Uttermost Part of the Earth,* passim, but especially its first half.

Essay 6 Tax and the Selfish Girl

1   See my article "The Oracles and their Cessation," *Encounter*, April 1989, reprinted in D. C. Stove, *Cricket Versus Republicanism* (Sydney: Quakers Hill Press, 1995).

2   David Hume, *A Treatise of Human Nature* (Oxford: Clarendon Press, 1988), Book I, Part II, section, ii, page 31. The italics are in the text.

3   Charles Darwin, *The Descent of Man, and Selection in Relation to Sex* (London: John Murray, 1874), Vol. I, pages 199ff.

4   M. T. Ghiselin, *The Economy of Nature and the Evolution of Sex* (Berkeley: University of California Press, 1974), page 247.

5   G. Hardin, *Sociobiology and Human Nature* (San Fransisco, Washington, and London: Jossey-Bass Publishers, 1978), pages 183–94.

6   Richard Dawkins, *The Selfish Gene* (New York: Oxford University Press, 1978), page 215 for the first part of this quotation, page 3 for the second.

7   Richard Dawkins, *The Extended Phenotype* (Oxford: W. H. Freeman and Co., 1982), page 57.

8   Ibid., page 56.

9   Ibid.

10   E. O. Wilson, *Sociobiology* (Cambridge: Harvard University Press, 1975), page 129.

11   Dawkins, *The Extended Phenotype*, page 57.

12   Dawkins, *The Selfish Gene*, page 110.

13   See for example, Wilson, *Sociobiology*, Chapter 5, and Dawkins, ibid., page 215.

14   See Dawkins, *The Extended Phenotype*, page 237.

15   Hardin, *Sociobiology and Human Nature*, page 194. The italics are not in the text.

16   Harriet Beecher Stowe, *Uncle Tom's Cabin* (Oxford: Oxford University

Press, 1998), page 55.

17 Hardin, *Sociobiology and Human Nature*, pages 188 and 190.

18 Ibid., page 188.

19 Joseph Butler, *Sermons* (London: Macmillan, 1913).

20 Mary Midgley, *Evolution as a Religion* (London: Methuen, 1985), page 126.

21 Bernard Mandeville, *The Fable of the Bees* (New York: Penguin Classics, 1970), page 87.

22 Ibid., page 81.

23 Ibid., page 88.

24 David Hume, *An Enquiry Concerning the Principles of Morals*, edited by D. Nidditch (Oxford: Oxford University Press, 1975), page 214.

25 Ibid.

26 See, for example, Robert Axelrod, *The Evolution of Cooperation* (New York: Basic Books, 1984).

27 See Derek Freeman, *Margaret Mead and Samoa* (Cambridge: Harvard University Press, 1983).

28 Colin Turnbull, *The Mountain People* (New York: Simon and Schuster, 1972).

29 See Dawkins, *The Selfish Gene*, page 205.

30 Ibid.

31 See T. H. Huxley, *Methods and Results* (London: Macmillan, 1985), page 306.

Essay 7 Genetic Calvinism, or Demons and Dawkins

1 Richard Dawkins, *The Selfish Gene* (New York: Oxford University Press, 1976).

2 See *William Bateson, F.R.S. Naturalist, His Essays and Addresses* (Cambridge: Cambridge University Press, 1928), for example, page 203.

3 See John Calvin, *Institutes of the Christian Religion* (1536), translated by H. Beveridge (Edinburgh: Edinburgh Printing Company, 1845), Vol. I, pages 203–9.

4 *The Selfish Gene*, page 210.

5 *The Selfish Gene*, page 211.

6 *The Selfish Gene*, page 95

7 *The Selfish Gene*, page 4. Italics in text.

8 *The Selfish Gene*, page 4. Italics in text.

9 *The Selfish Gene*, page 95. Italics in text.

10 *The Selfish Gene*, page 37. Italics in text.

11 *The Selfish Gene*, page 37.

12 Mary Midgley, "Gene-juggling," *Philosophy* 54, 1979, page 446. I am much indebted to this excellent article.

13 *The Selfish Gene*, page x.

14 *The Selfish Gene*, page 215. Italics not in text.
15 *The Selfish Gene*, page 3. Italics in text.
16 *The Selfish Gene*, page 215.
17 "Gene-juggling," page 456.
18 See my article "The Columbus Argument" in David Stove, *On Enlightenment*, edited by Andrew Irvine (New Brunswick: Transaction, 2003), pages 149–153.
19 Daniel C. Dennett, "Memes and the Exploitation of Imagination," *Journal of Aesthetics and Art Criticism*, 48:2, Spring 1990.
20 See *The Collins Encyclopedia of Animal Behaviour* (London: Collins, 1978), page 138.
21 See, for example, R. A. Fisher, *The Genetical Theory of Natural Selection* (1930) (New York: Dover Publications, 1958), pages 7–9.
22 *The Selfish Gene*, page 207. I have put square brackets around "Pythagoras's Theorem" to indicate that I have changed the example given in the text.
23 Richard Dawkins, *The Blind Watchmaker* (New York: W. W. Norton, 1986).
24 Richard Dawkins, *The Extended Phenotype: The Gene as the Unit of Selection* (Oxford and San Francisco: W. H. Freeman, 1982).
25 *The Extended Phenotype*, page 158. Italics not in text.
26 *The Blind Watchmaker*, page 126. Italics not in text.

Essay 8 Altruism and Shared Genes

1 William Hazlitt, *Selected Writings*, edited by Ronald Blyth (New York: Penguin Books, 1970), page 460.
2 Richard Dawkins, *The Selfish Gene* (New York: Oxford University Press, 1976), page 97.
3 W. D. Hamilton, "The Genetical Evolution of Social Behavior," *Journal of Theoretical Biology*, 7, 1964, page 19.
4 Richard D. Alexander, *Darwinism and Human Affairs* (Seattle: University of Washington Press, 1979); Michael Ruse, *Taking Darwin Seriously* (Oxford: Basil Blackwell, 1986); Robert Trivers, *Social Evolution* (Menlo Park, California: Benjamin/Cummings Publishing Co., 1985).
5 Charles Darwin, *The Origin of Species* (1859 facsimile reprint, Cambridge: Harvard University Press, 1966), page 81.
6 "The Genetical Evolution of Social Behavior."
7 See for example J. B. S. Haldane, *Possible Worlds* (London: Chatto and Windus, 1928), pages 32–3; and R. A. Fisher, *The Genetical Theory of Natural Selection* (New York: Dover Publications, 1958), pages 177–81.
8 On all of these cases, see for example Alexander, *Darwinism and Human Affairs*, pages 43–7.
9 *The Origin of Species*, page 236.

10 "The Genetical Evolution of Social Behavior," page 2. I have written "sibling" where Hamilton wrote "sib."

11 "The Genetical Evolution of Social Behavior," pages 25ff.

12 And compare the same author's article "Altruism and Related Phenomena, Mainly in Social Insects," in *Annual Review of Ecology and Systematics*, Vol. 3, 1972.

13 *The Selfish Gene*, page 102.

14 George C. Williams, *Adaptation and Natural Selection* (Princeton: Princeton University Press, 1974), page 195.

15 "The Genetical Evolution of Social Behavior," page 49.

16 *The Selfish Gene*, page 113.

17 "The Genetical Evolution of Social Behavior," page 16.

18 *The Selfish Gene*, page 113.

19 "The Genetical Evolution of Social Behavior," page 16. Note that I have inserted the words in square brackets.

20 Personal communication.

21 Mary Midgley, *Beast and Man* (Ithaca: Cornell University Press, 1978), pages 169–74.

22 *Adaptation and Natural Selection*, pages 197 and 252.

23 M. Roberts, *The Serpent's Fang* (London: Eveleigh Nash and Grayson, 1930), page 207.

24 E. O. Wilson, *Sociobiology: The New Synthesis* (Cambridge: Harvard University Press, 1975), page 3.

25 *Darwinism and Human Affairs*, page 46.

26 *The Selfish Gene*, page 95.

27 See, for example, Trivers, *Social Evolution*, page 109.

28 Francis Hutcheson, *An Inquiry Concerning the Original of Our Ideas of Virtue or Moral Good* (the 1897 edition), reprinted in part in L. A. Selby-Bigge, editor, *British Moralists* (Oxford: Oxford University Press, 1987), Vol. I, page 95. I am indebted for this reference to my old friend Dr. Eric Dowling.

29 The sources of the first two of these quotations were given in Essay 6, Notes 4 and 5 respectively. The source of the third quotation is Alexander, *Darwinism and Human Affairs*, page 133.

30 *The Selfish Gene*, page 215.

31 Walt Kelly, *Ten Ever-Lovin' Blue-Eyed Years with Pogo* (New York: Simon and Schuster, 1959), page 170.

Essay 9   A New Religion

1 Richard Dawkins, *The Selfish Gene* (New York: Oxford University Press, 1976), page x.

2 Ibid., p. 185.

3 Richard Dawkins, *The Extended Phenotype* (Oxford: Freeman & Co.,

1982), page 158.

4   Richard Dawkins, *The Blind Watchmaker* (London: Longman, 1986), page 126.

5   E. O. Wilson, *Sociobiology: The New Synthesis* (Cambridge: Harvard University Press, 1975), page 3.

6   *The Selfish Gene*, page 210.

7   See Dawkins, *The Extended Phenotype*, page 57.

8   Ibid., Chapter 4.

9   Ibid., Chapter 13.

10   Ibid., page 200.

11   Ibid., pages 68–70.

12   Ibid., page 59.

13   *The Selfish Gene*, page 36.

14   *The Extended Phenotype*, page 164.

15   *The Selfish Gene*, page 36.

Essay 10 Paley's Revenge, or Purpose Regained

1   George C. Williams, *Adaptation and Natural Selection* (Princeton: Princeton University Press, 1974).

2   Ibid., page v.

3   Ibid., pages 97, 194, 196, 207.

4   See, e.g., ibid., pages 197, 199 and especially 252.

5   Richard Dawkins, *The Selfish Gene* (New York: Oxford University Press, 1976), page 12.

6   For example, E. Sober, *The Nature of Selection* (Cambridge: MIT Press, 1985), "Acknowledgements."

7   David Hume, *Dialogues Concerning Natural Religion*, edited by Richard Wollheim (London: William Collins Sons/Fontana Library, 1963), page 128. The italics are not in the text.

8   From a "Notice" prefixed to Sir Charles Bell, *The Hand*, 9th edition (London: George Bell and Son, 1874).

9   See Williams, *Adaptation and Natural Selection*, page 54.

10   The first of these passages is from Francis Darwin, editor, *The Life and Letters of Charles Darwin* (London: John Murray, 1888), Vol. I, page 47; the second passage is from ibid., Vol. II, page 219.

11   Richard Dawkins, *The Blind Watchmaker* (New York: W. W. Norton, 1986), page 5.

12   Williams, *Adaptation and Natural Selection*, page 133.

13   Ibid., page 261. Italics not in text.

14   Ibid., page 209. Italics not in text.

15   Ibid., page 212. Italics not in text.

16   Ibid., page 160. Italics not in text.

17   Ibid., pages 96–7. Italics not in text.

18 Ibid., page 256. Italics not in text.
19 Ibid., page 68. Italics not in text.
20 Ibid., page 189. Italics not in text.
21 Ibid., page 44. Italics not in text.
22 Ibid., page 70. Italics not in text.
23 Ibid., pages 258–9.
24 See ibid., pages 161, 194.
25 Ibid., page 193. Italics not in text.
26 *The Blind Watchmaker*, page 5.
27 *The Selfish Gene*, page 25.
28 See, for example, ibid., page 38.
29 Ibid., page 95.
30 Ernest Nagel, *The Structure of Science* (London: Routledge and Kegan Paul, 1961).
31 Andrew Woodfield, *Teleology* (Cambridge: Cambridge University Press, 1976).
32 Alan Olding, *Modern Biology and Natural Theology* (London: Routledge, 1991).
33 William Paley, *Natural Theology* (London: Longman and Co., 1838), pages 20 and 186.
34 Charles Darwin, *On the Origin of Species* (Cambridge: Harvard University Press, 1996), pages 78–9.
35 Ibid., pages 66–7.
36 See, for example, ibid., pages 102 and 113.
37 Paley, *Natural Theology*, page 175.
38 Ibid., page 235.
39 Ibid., page 247.
40 Ibid., page 244.
41 See Arthur Schopenhauer, *The World as Will and Representation*, translated by E. Payne (New York: Dover Publications, 1969), Index, s.c. "Youth."
42 See H. O. Taylor, *Greek Biology and Medicine* (London: Harrap, 1923), pages 81–2.
43 Paley, *Natural Theology*, page 222; cf. page 263.
44 See Desmond King-Hele, *Erasmus Darwin* (London: Macmillan, London, 1963), pages 135–7.
45 Some information about the history of contraception can be found in my articles in *Encounter* in May and June 1990.
46 See Jonathan Beecher and Richard Bienvenu *The Utopian Vision of Charles Fourier* (London: Jonathan Cape, 1975), e.g., pages 381–95.
47 Hume, *Dialogues Concerning Natural Religion*, page 172.
48 Ibid., page 186.
49 Arthur Schopenhauer, *The World as Will and Representation*, *passim*, but more especially in the second volume of that book.
50 See, for example, ibid., Vol. II, pages 312 and 396.

51 See, for example, ibid., Vol. II, page 209ff.
52 See, for example, ibid., Vol. II, page 331ff.
53 See ibid., Index, s. v., "Owen, R." and "Kirby and Spence."
54 See, for example, ibid., Vol. II, pages 338 and 506.
55 See, for example, ibid., Vol. II, page 336.
56 Ibid., Vol. II, page 356.
57 David Stove, *The Plato Cult, and Other Philosophical Follies* (Oxford: Basil Blackwell, 1991) pages 19–21.
58 Arthur Schopenhauer, *The World as Will and Representation*, Vol. I, the first book.
59 See Stove, *The Plato Cult*, Essay 6.
60 *Adaptation and Natural Selection*, page 255.
61 Charles Darwin, *On the Origin of Species*, pages 66–7.
62 See Gavin De Beer, *Streams of Culture* (New York: Lippincott, 1969), page 163.
63 A. O. Lovejoy, *The Great Chain of Being* (New York: Harper Torchbooks, 1960).
64 I borrow this useful phrase from the title of the book by E. J. Dijksterhuis, *The Mechanization of the World Picture* (New York: Oxford University Press, 1961).
65 R. A. Fisher, *The Genetical Theory of Natural Selection* (New York: Dover Publications, 1958), chapters X–XII.
66 Cf. Essay 11 below.
67 See Arthur Schopenhauer, *The World as Will and Representation*, Vol. II, page 358.

Essay 11 Errors of Heredity

1 Francis Galton, *Hereditary Genius*, edited by C. D. Darlington (London: Macmillan, 1892), page 13.
2 E. O. Wilson, *Sociobiology: The New Synthesis* (Cambridge: Harvard University Press, 1975), page 129.
3 Richard Dawkins, *The Selfish Gene* (New York: Oxford University Press, 1976), page 110. Italics in text.
4 Charles Darwin, *The Descent of Man, and Selection in Relation to Sex* (London: John Murray, 1871), Vol. I, pages 69–70.
5 See R. A. Fisher, *The Genetical Theory of Natural Selection* (New York: Dover Publications, 1930), especially Chapter XI.
6 Ibid., page 265.
7 See, for example, C. G. Darwin, *The Next Million Years* (London: Rupert Hart-Davis, 1952), page 183.
8 See Galton, *Hereditary Genius*, pages 410–2.
9 See Gwen Raverat, *Period Piece*, (London: Faber and Faber, 1952), Chapter VII, especially pages 133–4.

10   See Samuel Butler, *Erewhon* (London: Jonathan Cape, 1872), pages 118–24.

11   R. D. Alexander, *Darwinism and Human Affairs* (Seattle: University of Washington Press, 1979), page 47.

12   Charles Darwin, *On the Origin of Species* (Cambridge: Harvard University Press, 1966), pages 80–1.

13   See R. Trivers, *Social Evolution* (Menlo Park: Benjamin/Cummings Publishing Co., 1985), pages 198–9.

14   Raverat, *Period Piece*, page 209.

Index